VEDIC MATHEMATICS TEACHER'S MANUAL

INTERMEDIATE LEVEL

Kenneth R. Williams

INSPIRATION BOOKS

Published by Inspiration Books, 2002.
Kensglen, Nr Carsphairn, Castle Douglas, DG7 3TE, Scotland, U.K.

ISBN 978-1-902517-17-9

Revised 2009

www.vedicmaths.org

PREFACE

This Manual is the second of three self-contained Manuals (Elementary, Intermediate and Advanced) which are designed for adults with a basic understanding of mathematics to learn or teach the Vedic system. So teachers could use it to learn Vedic Mathematics, though it is not suitable as a text for children (for that the Cosmic Calculator Course is recommended). Or it could be used to teach a course on Vedic Mathematics. This Manual is suitable for teachers of children of grades 5 to 10.

The sixteen lessons of this course are based on a series of one week summer courses given at Oxford University by the author to Swedish mathematics teachers between 1990 and 1995. Those courses were quite intensive consisting of eighteen, one and a half hour, lessons.

The lessons in this book however probably contain more material than could be given in a one and a half hour lesson. The teacher/reader may wish to omit some sections, go through the material in a different sequence to that shown here or break up some sections (e.g. recurring decimals).

All techniques are fully explained and proofs are given where appropriate, the relevant Sutras are indicated throughout (these are listed at the end of the Manual) and, for convenience, answers are given after each exercise. Cross-references are given showing what alternative topics may be continued with at certain points.

It should also be noted that in the Vedic system a mental approach is preferred so we always encourage students to work mentally as long as it is comfortable. In the Cosmic Calculator Course pupils are given a short mental test at the start of most or all lessons, which makes a good start to the lesson, revises previous work and introduces some of the ideas needed in the current lesson. In the Vedic system pupils are encouraged to be creative and use whatever method they like.

Some topics will be found to be missing in this text: for example, there is no section on area, only a brief mention. This is because the actual methods are the same as currently taught so that the only difference would be to give the relevant Sutra(s).

CONTENTS

CONTENTS

LESSON 1
BASIC DEVICES

SUMMARY

1.1 **Introduction** - background information about Vedic Mathematics.
1.2 **Digit Sums** – reducing a number to a single digit, and the 9-point circle.
1.3 **Left to Right** – mental addition and multiplication from left to right.
1.4 **Writing Left to Right Sums**
1.5 **Checking Devices** – checking multiplication and addition using the digit sums and by examining the first and last figures.
1.6 **Subtraction** – from left to right and checking.

1.1 INTRODUCTION

Vedic Mathematics is the ancient system of mathematics which was rediscovered early last century by **Sri Bharati Krsna Tirthaji** (henceforth referred to as Bharati Krsna).

The Sanskrit word "veda" means "knowledge". The Vedas are ancient writings whose date is disputed but which date from at least several centuries BC. According to Indian tradition the content of the Vedas was known long before writing was invented and was freely available to everyone. It was passed on by word of mouth. The writings called the Vedas consist of a huge number of documents (there are said to be millions of such documents in India, many of which have not yet been translated) and these have recently been shown to be highly structured, both within themselves and in relation to each other (see Reference 2). Subjects covered in the Vedas include Grammar, Astronomy, Architecture, Psychology, Philosophy, Archery etc., etc.

A hundred years ago Sanskrit scholars were translating the Vedic documents and were surprised at the depth and breadth of knowledge contained in them. But some documents headed "Ganita Sutras", which means mathematics, could not be interpreted by them in terms of mathematics. One verse, for example, said "in the reign of King Kamse famine, pestilence and unsanitary conditions prevailed". This is not mathematics they said, but nonsense.

Bharati Krsna was born in 1884 and died in 1960. He was a brilliant student, obtaining the highest honours in all the subjects he studied, including Sanskrit, Philosophy, English, Mathematics, History and Science. When he heard what the European scholars were saying about the parts of the Vedas which were supposed to contain mathematics he resolved to study the documents and find their meaning. Between 1911 and 1918 he was able to reconstruct the ancient system of mathematics which we now call Vedic Mathematics.

He wrote sixteen books expounding this system, but unfortunately these have been lost and when the loss was confirmed in 1958 Bharati Krsna wrote a single introductory book entitled "Vedic Mathematics". This is currently available and is a best-seller (see Reference 1).

The present author came across the book "Vedic Mathematics" in 1971 and has been developing the content of that book, and applying the system in other areas not covered by Bharati Krsna, since then. Anything in this book which is not in "Vedic Mathematics" has been developed independently by the author in this way.

There are many special aspects and features of Vedic Mathematics which are better discussed as we go along rather than now because you will need to see the system in action to appreciate it fully. But the main points for now are:

1) The system rediscovered by Bharati Krsna is based on sixteen formulae (or Sutras) and some sub-formulae (sub-Sutras). These Sutras are given in word form: for example *Vertically and Crosswise* and *By One More than the One Before*. In this text they are indicated by italics. The Sutras can be related to natural mental functions such as completing a whole, noticing analogies, generalisation and so on.

2) Not only does the system give many striking general and special methods, previously unknown to modern mathematics, but it is far more coherent and integrated as a system.

3) Vedic Mathematics is a system of mental mathematics (though it can also be written down).

Many of the Vedic methods are new, simple and striking. They are also beautifully interrelated so that division, for example, can be seen as an easy reversal of the simple multiplication method (similarly with squaring and square roots). This is in complete contrast to the modern system. Because the Vedic methods are so different to the conventional methods, and also to gain familiarity with the Vedic system, it is best to practice the techniques as you go along.

"The Sutras (aphorisms) apply to and cover each and every part of each and every chapter of each and every branch of mathematics (including arithmetic, algebra, geometry – plane and solid, trigonometry – plane and spherical, conics-geometrical and analytical, astronomy, calculus – differential and integral etc., etc. In fact, there is no part of mathematics, pure or applied, which is beyond their jurisdiction"
From "Vedic Mathematics", Page xvi.

1.2 DIGIT SUMS

The **digit sum** of a number is found by adding the digits in a number and adding again if necessary until a single figure is reached.

This is useful for divisibility testing, and, for our present purposes, for checking calculations.

1 So, for example, the digit sum of **23** is 5 as 2+3 = **5**.

Similarly for **21302** we get 2+1+3+0+2 = **8**.

And for **76** we first get 7+6 = 13, but 1+3 = 4, so 76 → **4**.

For **123456** we get 21 which gives 3, so 123456 → **3**.

So every number, no matter how long, can be reduced to a single figure, its digit sum.

Practice A Find the digit sum of the following numbers:

a 42 **b** 47 **c** 32101

d 777 **e** 2468 **f** 991 **g** 837364

---

a 6 **b** 2 **c** 7 **d** 3 **e** 2 **f** 1 **g** 4

We can understand how these digit sums work by using the nine point circle in which the numbers from 1 to 9 are drawn round a circle as shown below.

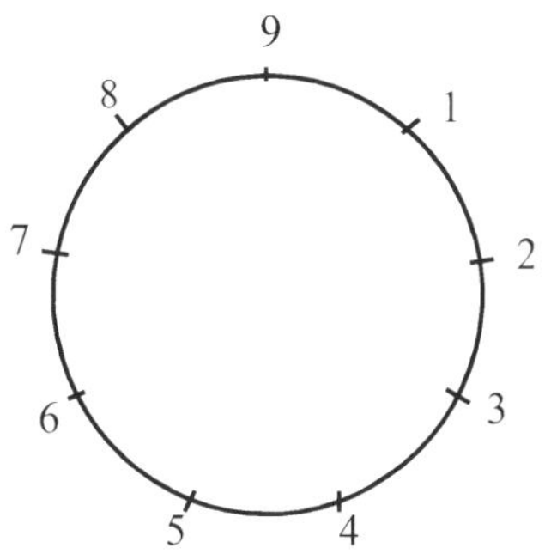

This circle is useful in a number of ways, as we will see.

Continuing to number round the circle it seems reasonable to put 10 after the 9 and then 11 at the same place as 2, and so on:

So all whole numbers find their place on the circle.

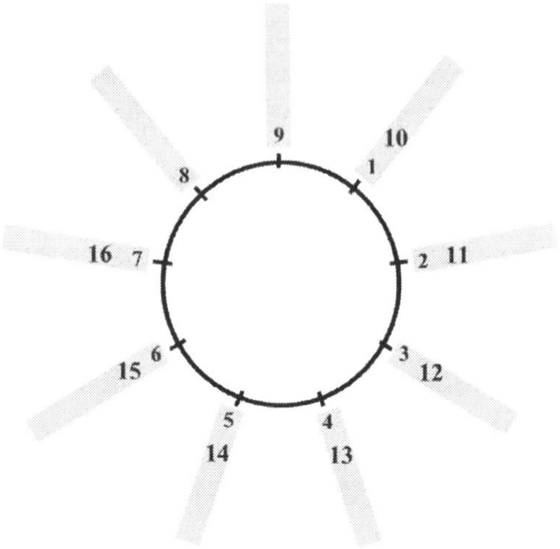

Now if we add the digits of 10 we get 1, which is the number beside it. And if we add the digits of 11 we get 2 which is the number beside it, and so on.

This means that all numbers with a digit sum of 2 will be on the 2-branch, and similarly for the other branches.

This also shows that adding 9 to a number does not affect its digit sum: whatever branch a number is on, adding 9 (or any number of 9's, or subtracting 9's) to it will not change the branch it is on.

And this means that if we have to find the digit sum of a number with a nine in it **we can delete the 9**.

2 The digit sum of **49** for example is just **4** (because 49 and 40 will be on the same branch): we simply delete the 9.

Similarly for the digit sum of **39979** we ignore the 9's, add the 3 and 7 and give the digit sum as **1** (10→1).

This can be taken further: **any pair or group of digits which add up to 9 can be deleted.**

3 For **3725** we can delete the 72 in the centre because they add up to 9. We just add 3 and 5 to get the digit sum of **8**.

For **613954** we see 6 and 3 which make 9, 5 and 4 which make 9 and also a 9. Deleting all these we have only 1 left so 613954 → **1**.

The number **3251** also contains a 9 because 3+5+1 = 9, so deleting these 3251 → **2**.

✎ **Practice B** Find the digit sum:

a 392	**b** 93949	**c** 789789	**d** 456	**e** 81187	**f** 62371
g 59479	**h** 1999	**i** 480417	**j** 13579	**k** 24680	**l** 123456789

a 5	**b** 7	**c** 3	**d** 6	**e** 7	**f** 1
g 7	**h** 1	**i** 6	**j** 7	**k** 2	**l** 9

The digit sum for 5454 is 9 by adding the digits, but if we delete the 9's (because 5+4 = 9) we get 0 for the digit sum.

In fact in digit sums 9 and 0 are equivalent and this can also be seen on the 9-point circle: since 0 is the number before 1 we would expect it to be on the same branch as 9 (counting backwards round the circle: . . 3, 2, 1, 0 . . .). So we can say 5454 → 9 or 5454 → 0; both are correct. This will be significant when we are checking subtraction sums in Section 1.5. You can even include negative numbers on the circle by numbering backwards round the circle.

The Sutra *When the Samuccaya is the Same it is Zero* was in fact in use when we were casting out nines or figures that total nine in Lesson 1. For example, to get the digit sum of 425 we can cast out the 4 and 5, as they total nine, and give the digit sum as 2: *when the total is the same (as 9) it is zero (can be cast out)*. (The term *Samuccaya* means combination or total.)

The same Sutra also applies in cancelling a common factor from the top and bottom of a fraction.

1.3 LEFT TO RIGHT

1.3a ADDITION

The Vedic system is extremely flexible and teaches flexibility. Conventional addition of, say, 2-figure numbers works from right to left, but since we write and pronounce numbers from left to right it is not so easy to add numbers mentally this way. Vedic addition includes addition from left to right. You may prefer to do Section 1.4 before this section as it shows **written** calculations from left to right.

4 Given 76
88 +
__ for example, we can see that we have 7 + 8 = 15 in the left column.

And a glance at the right-hand column shows that there is a carry and so the first two figures of the answer must be 16. The right-hand total is 14 and since the 1 has been dealt with already we only have to put the 4 after the 16 to get 164 as the answer.

The totals 15 and 14 are combined as shown below:

```
1 5
  1 4 +
1 6 4
```

This is an easy, direct and natural way of adding numbers.

5

```
  6 4 6
  6 7 8 +
1 3 2 4
```

The first total is 12 which becomes 13 because there is a carry in the next column. The 11 in the second column means that we now have 131. Then remembering this 131 we look at the third column which adds up to 14. The 131 therefore becomes 132 and a 4 is placed at the end giving 1324 as the answer.

Mental mathematics obviously relies more on the memory than conventional methods where every step is written down. Young children have very good memories and mental mathematics helps to strengthen the memory further. (This means that Vedic Mathematics is good for adults too, whose memory may not be so good.) This also gives confidence and teaches self-reliance, showing that we do not need pencil and paper or calculator for every sum but can find an answer without any external help.

The totals which are combined in the above sum can be shown as:

```
1 2
  1 1
    1 4 +
1 3 2 4
```

This shows nicely how the column totals are combined but if we are encouraging a mental system of calculation we do not want to encourage children to write this out but to build the answer up step by step, mentally, until it is complete, and then, if required, write the answer down.

We can show the actual steps more clearly as follows:

```
first   1 2            then   1 3 1
          1 1 +                   1 4 +
        1 3 1                 1 3 2 4
```

Or, even more compactly as: 12,11 = 131;

131,14 = **1324**.

Here the curved line under two numbers shows that they are to be added. This is the method we will use in this book to indicate which figures are to be mentally combined.

6 Similarly for

```
8 6 5 5
8 9 2 8 +
_______
```

the totals give 16,15 = 175,

then 175,7 = 1757 (there is no carry here),

then 1757,13 = **17583**.

Practice C Add the following, writing down only the answer:

a	**b**	**c**	**d**	**e**	**f**
8 6	4 7	7 3	3 7	6 7 8	8 3 6
7 7	8 8	6 4	4 9	7 8 7	6 2 7
___	___	___	___	___	___

It is also useful to practice adding sums after only hearing the numbers being added, for example:

g	**h**	**i**	**j**	**k**	**l**
7 7	6 7	3 8	8 2	7 7 7	2 8 8
8 8	6 8	4 8	8 3	8 8 8	2 8 9
___	___	___	___	___	___

a	163	b	135	c	137	d	86	e	1465	f	1463
g	165	h	135	i	86	j	165	k	1665	l	577

1.3b MULTIPLICATION

The same idea can be used for multiplication.

7 Suppose we have the sum:

$$\begin{array}{r} 2\ 3\ 7 \\ \underline{2}\ \times \\ \overline{\quad\quad} \end{array}$$

We multiply each of the figures in 237 by 2 starting at the left.
The answers we get are **4, 6, 14**.

Since the 14 has two figures the 1 must be carried leftwards to the 6.
So, $4\underset{\smile}{6,1}4$ = **474**.

Again we build the answer up mentally from the left:
first 4, then 4,6 = 46, then $4\underset{\smile}{6,1}4$ = **474**.

8

$$\begin{array}{r} 2\ 3\ 6 \\ \underline{7}\ \times \\ \overline{\quad\quad} \end{array}$$

First we have 14,
then $1\underset{\smile}{4,2}1$ = 161.

then $16\underset{\smile}{1,4}2$ = **1652**.

9 For 73 × 7 we get $4\underset{\smile}{9,2}1$ = **511**. (because 49+2 = 51)

"Little boys come dancing forward with joy and professors ask, 'well, how can the answer be written down without any intermediate steps of working at all?'".
From "Vedic Metaphysics", Page 168.

✐ **Practice D** Multiply the following from left to right:

a 27 × 3	**b** 72 × 7	**c** 26 × 6	**d** 76 × 6	**e** 78 × 9	**f** 83 × 3
g 642 × 4	**h** 256 × 3	**i** 741 × 3	**j** 223 × 9		
k 1059 × 7	**l** 8631 × 4	**m** 5432 × 8	**n** 4097 × 7		

---

a 81	b 504	c 156	d 456	e 702	f 249
g 2568	h 768	i 2223	j 2007		
k 7413	l 34524	m 43456	n 28679		

These can also be practised when the sum is heard and not seen.

✐ **Practice E** Multiply mentally from the left [say the sums rather than write them up]:

a 33 × 4	**b** 55 × 7	**c** 34 × 8	**d** 66 × 6	**e** 62 × 4
f 55 × 5	**g** 44 × 4	**h** 43 × 7	**i** 88 × 4	**j** 73 × 3

---

a 132	b 385	c 272	d 396	e 248
f 275	g 176	h 301	i 352	j 219

1.3c ADVANTAGES OF LEFT TO RIGHT CALCULATION

There are many advantages to left to right calculation as we pronounce and write numbers from left to right. Also, sometimes we only need the first two or three significant figures and would waste a lot of time and effort if we found all the figures of a long sum by starting at the right. Division is always done from the left, so all calculations can be done left to right, which means we can combine operations and, for example, find the square root of the sum of two squares in one line (see Chapter 16). For finding square roots, trig functions and so on there is no right-hand figure to start from anyway, so there is no option but to start at the left.

1.4 WRITING LEFT TO RIGHT SUMS

We have described the mental method of left to right addition and multiplication. Here we see how to proceed if the sums are written out instead. You may prefer to do this Section before the mental method: Section 1.3.

(10) **367 + 985 = 1352**.

$$\begin{array}{r} 3\,6\,7 \\ 9\,8\,5 + \\ \hline 1_2 \quad\quad \end{array}$$

$$\begin{array}{r} 3\,6\,7 \\ 9\,8\,5 + \\ \hline 1_2 3_4 5\,2 \end{array}$$

In this addition sum, if we add the left-hand column we get 3 + 9 = 12. Since there will be a carried figure from the next column that will affect this total, we put down only the 1 and carry the 2 as forwards as shown. The middle column adds up to 14, We add the carried 2, **as 20**, to this to get 34 and put down 3_4 as shown. Then the right-hand column adds up to 12, to which we add the carried 4, **as 40**, to get 52: which goes down to complete the answer.

(11) **3457 × 8 = 27656**.

$$\begin{array}{r} 3\;\;4\;\;5\;\;7 \\ 8 \times \\ \hline 2\,{}_4 7\,{}_2 6\,{}_0 5\;\;6 \end{array}$$

We begin on the left: 3 × 8 = 24, put as shown;
4 × 8 = 32, add the carried 4, **as 40**, 32 + 40 = 72;
5 × 8 = 40, add the carried 2, **as 20**, 40 + 20 = 60;
7 × 8 = 56, 56 + 0 = 56, put as shown.

(12) **138 × 4** = $0_4 5_2 52$ = **552**.

In the previous two examples we started with a 2-figure product (12 and 24).
Here we initially get 1 × 4 = 4, a single figure number. So we put down 0_4 as shown.

(13) **234 × 3** = $0_6 6_9 {}^1 02$ = **702**.

Here we get 0_6 and then 6_9 and then we add 12 to 90 to get 102.
As this is a 3-figure number the 1 in 102 is carried over to the 6, so that it becomes 7.

✎ **Practice F** Try the sums in Practice C and D from left to right like this.

General multiplication and squaring in one line, from left to right (or from right to left) are covered in Lessons 6 and 7.

1.5 CHECKING DEVICES

There are various checking devices available in the Vedic system.

A One comes under the Sutra *The First by the First and the Last by the Last* (see the list of Sutras on Pages 194,5).

Suppose we have found that $877 \times 3 = 2631$ and we want to check the answer. The above Sutra allows us to check the beginning and end of the answer.

Multiplying the first figure of 877 by the first figure of 3 we get $8 \times 3 = 24$ and our answer starts with 26, so this looks right. Then multiplying the last figure of 877 by the last figure of 3 we get $7 \times 3 = 21$, telling us that the answer must end with a 1, which it does.

This is a useful and very quick check. It does not tell us that the answer is right of course but it could tell us that the answer is wrong.

The Sutra (in fact it is a sub-Sutra) *The First by the First and the Last by the Last* is used in many ways. For example in measuring or drawing a line with a ruler (or an angle with a protractor) we line the first point of the line with the first mark on the ruler and note the position of the last point on the ruler.

B The Digit Sum Check.

The digit sums can be used to check sums. We simply replace all the numbers in the sum by their digit sums and check that the calculation is still true.

14 To check **877 × 3 = 2631** we replace 877 by its digit sum which is 4.
And similarly 2631 is replaced by 3.

Then $877 \times 3 = 2631$ becomes $\mathbf{4 \times 3 = 3}$,

and this is true in digit sum arithmetic as $4 \times 3 = 12$ and 12 is equivalent to 3:
($12 \rightarrow 1+2 = 3$).

The Vedic formula here is *The Product of the Sum is the Sum of the Products*.

It is worth while doing the check next to the sum:

$$\begin{array}{r} 8\ 7\ 7 \\ \underline{3}\ \times \\ \underline{2\ 6\ 3\ 1} \end{array} \qquad \begin{array}{r} 4 \\ \underline{3}\ \times \\ \underline{3} \end{array}$$

So we write the three digit sums to the right of the sum and check it forms a correct sum.

✏ **Practice G** Use the methods above to check your answers to Practice D.

15 Checking addition is also very easy

8 7 5		2			7 3 5	6
6 4 6 +		7 +	and		8 8 +	7 +
1 5 2 1		9			8 2 3	4

Note: a) because these are addition sums we **add** the digit sums (not multiply as before),
b) the second sum is verified because 6+7 = 13 → 4 in digit sums.

Practice H Check your answers to Practice C using the digit sums.

The digit sum check does not prove the answer is correct: if you add 44 and 77 and give the answer as 211 instead of 121 the digit sum check will not detect it. The checks show that the answer is probably correct.
The sums can also be checked of course by doing the calculation from right to left instead of from left to right.

The Vedic formula *The Product of the Sum is the Sum of the Products* applies for all the digit sum checks. For addition it would be *The Total of the Digit Sums is the Digit Sum of the Total*. The formula has many other applications (see Reference 3), for example in finding areas of composite shapes (*The Area of the Whole is the Sum of the Areas*).

1.6 SUBTRACTION

1.6a SUBTRACTION FROM LEFT TO RIGHT

This is also very easy.

16 Find **35567 – 11828**.

```
3 5 5 6 7
1 1 8 2 8 –
2
```

We set the sum out as usual:
Then starting on the **left** we subtract in each column.
3 – 1 = **2**, but before we put **2** down we check that in the next column the top number is larger. In this case 5 is larger than 1 so we put **2** down.

```
3 5 ¹5 6 7
1 1 8 2 8 –
2 3
```

In the next column we have 5 – 1 = 4, but looking in the third column we see the top number is not larger than the bottom (5 is less than 8) so instead of putting 4 down we put **3** and the other 1 is placed *On the Flag,* as shown so that the 5 becomes 15.

So now we have 15 – 8 = **7**. Checking in the next column we can put this down because 6 is greater than 2.
In the fourth column we have 6 – 2 = 4, but looking at the next column (7 is smaller than 8) we put down only **3** and put the other one *On the Flag* with the 7 as shown.

$$\begin{array}{r} 3\ 5\ {}^{1}5\ 6\ {}^{1}7 \\ \underline{1\ 1\ 8\ 2\ 8} - \\ \mathbf{2\ 3\ 7\ 3} \end{array}$$

Finally 17 – 8 = **9**:

$$\begin{array}{r} 3\ 5\ {}^{1}5\ 6\ {}^{1}7 \\ \underline{1\ 1\ 8\ 2\ 8} - \\ \mathbf{2\ 3\ 7\ 3\ 9} \end{array}$$

17 Find **535 – 138**.

Here we have 5 – 1 = 4 in the first column.
But in the next column the figures are the same.
In such a case we must look one further column along.

$$\begin{array}{r} 5\ 3\ 5 \\ \underline{1\ 3\ 8} - \end{array}$$

And since in the third column the top number is smaller, we reduce the 4 to **3**:

$$\begin{array}{r} 5\ {}^{1}3\ 5 \\ \underline{1\ 3\ 8} - \\ \mathbf{3} \end{array}$$

Then we proceed as before:

$$\begin{array}{r} 5\ {}^{1}3\ {}^{1}5 \\ \underline{1\ 3\ 8} - \\ \mathbf{3\ 9\ 7} \end{array}$$

> We subtract in each column starting on the left, but before we put an answer down we look in the next column.
> If the top is greater than the bottom we put the figure down.
> If not, we reduce the figure by 1, put that down and give the other 1 to the smaller number at the top of the next column.
> If the figures are the same we look at the next column to decide whether to reduce or not.

Practice I Subtract the following from left to right.

a 444 – 183	**b** 63 – 28	**c** 813 – 345	**d** 695 – 368	**e** 765 – 369	**f** 504 – 275
g 51 – 38	**h** 3456 – 281	**i** 7117 – 1771	**j** 8008 – 3839	**k** 5161 – 1838	**l** 9876 – 6789

a 261 (3–3=0=9) **b** 35 (9–1=8) **c** 468 (3–3=0=9) **d** 327 (11–8=3) **e** 396 (9–9=0=9) **f** 229 (9–5=4)
g 13 (6–2=4) **h** 3175 (9–2=7) **i** 5346 (7–7=0=9) **j** 4169 (7–5=2) **k** 3323 (4–2=2) **l** 3087 (3–3=0=9)

1.6b CHECKING SUBTRACTION SUMS

18 Find **69 – 23** and check the answer.

6 9	6	The answer is 46.
2 3 –	5 –	The digit sums of 69 and 23 are 6 and 5.
4 6	1	And 6 – 5 = 1, which confirms the answer because the digit sum of the answer, 46, is also 1.

19 Find **56 – 29**.

5 6	2
2 9 –	2 –
2 7	0

In this example, the digit sum of both 56 and 29 is 2 and subtracting gives us 0.
The digit sum of the answer is 9, but we have already seen that 9 and 0 are equivalent as digit sums [see end of Section 1.2], so the answer is confirmed.
Alternatively, we can add 9 to the upper 2 before subtracting the other 2 from it: 11–2 = 9. This is because adding 9 to any number takes you round the circle and back to where you started: 2 and 11 are equivalent in terms of digit sums.

20 Find **679 – 233**.

6 7 9	4
2 3 3 –	8 –
4 4 6	5

Here we have 4 – 8 in the digit sum check so we simply add 9 to the upper figure (the 4) and continue: 13 – 8 = 5, which is also the digit sum of 446.

✎ Practice J

Check your answers to Exercise I by using the digit sum check.

Answers: **see answers to previous exercise.**

LESSON 2
MORE BASIC DEVICES

SUMMARY

2.1 **Number Splitting** – simplifying mental calculations:
Addition/Subtraction/Multiplication/Division

2.2 ***All from 9 and the Last from 10*** – for subtractions and the use of bar numbers.

2.1 NUMBER SPLITTING

This is a very useful device for splitting a difficult sum into two or more easy ones and comes under the formula *By Alternate Elimination and Retention*.
For quick mental sums number splitting can considerably reduce the work involved in a calculation.

ADDITION

1 Suppose you are given the addition sum:

$$\begin{array}{r} \mathbf{2\ 3\ 4\ 5} \\ \underline{\mathbf{6\ 7\ 3\ 8}}\ + \\ \underline{} \end{array}$$

With 4-figure numbers it looks rather hard.
But if you split the sum into two parts, each part can be done easily and mentally:

$$\begin{array}{cc|cc} 2 & 3 & 4 & 5 \\ 6 & 7 & 3 & 8 \;+ \\ \hline \mathbf{9} & \mathbf{0} & \mathbf{8} & \mathbf{3} \end{array}$$

On the right we have 45 + 38 which (mentally) is **83**.
So we put this down.
And on the left we have 23 + 67 which is **90**.

2 Find

$$\begin{array}{r} \mathbf{6\ 3\ 4\ 3} \\ \underline{\mathbf{2\ 3\ 8\ 3}}\ + \\ \underline{} \end{array}$$

To avoid carries it is best to imagine two lines:

$$\begin{array}{c|cc|c} 6 & 3 & 4 & 3 \\ 2 & 3 & 8 & 3 \;+ \\ \hline \mathbf{8} & 7 & 2 & \mathbf{6} \end{array}$$

Practice A Add the following:

a	3 4 5 6 4 7 1 7	**b**	1 8 1 9 1 7 1 6	**c**	6 4 4 6 2 8 3 8	**d**	8 3 2 1 1 8 2 3
e	5 5. 1 6 6. 2	**f**	4 .5 5 4 3 .6 3 6	**g**	1 2 3 4 4 9 4 4	**h**	5 2 3 4 9 3 9 3

a	**81/73**	b	**35/35**	c	**92/84**	d	**101/44**
e	**121/.3**	f	**8.1/90**	g	**61/78**	h	**14/62/7**

SUBTRACTION

3 Suppose we have the subtraction sum:

5 4 5 4
1 7 2 6 –

We can split this up into two easy ones:

5 4	5 4
1 7	2 6 –
3 7	2 8

First 54 – 26 = **28**,
then 54 – 17 = **37**.

4

4 4 6 8
2 2 8 6 –

Splitting this in the middle like the last one would involve 68–86 which is not easy. So we split as follows:

4	4 6	8
2	2 8	6 –
2	1 8	2

Practice B Subtract the following:

a	3 2 4 3 1 3 1 9	**b**	4 4 4 4 1 8 2 8	**c**	7 0 7 0 1 5 2 6	**d**	3 7 2 1 1 9 0 9
e	6 8 8 9 1 9 3 6	**f**	8 5 2 1 3 9	**g**	7 7 7 5 8 5	**h**	6 6 6 6 2 9 3 8

a	**19/24**	b	**26/16**	c	**55/44**	d	**18/12**
e	**49/53**	f	**7/13**	g	**19/2**	h	**37/28**

MULTIPLICATION

The same technique can be applied here.

5 Find **352 × 2**

We can split this sum: 35 / 2 × 2 = **704**.
35 and 2 are easy to double.

6 Similarly **827 × 2** becomes 8 / 27 × 2 = **1654**,

604 × 7 becomes 6 / 04 × 7 = **4228**,

3131 × 5 becomes 3 / 13 / 1 × 5 = **15655**.

Practice C Multiply the Following:

a 432 × 3	**b** 453 × 2	**c** 626 × 2	**d** 433 × 3
e 308 × 6	**f** 814 × 4	**g** 515 × 5	**h** 919 × 3
i 1416 × 4	**j** 2728 × 2	**k** 3193 × 3	**l** 131415 × 3

---

a	**12/96**	b	**90/6**	c	**12/52**	d	**12/99**
e	**18/48**	f	**32/56**	g	**25/75**	h	**27/57**
i	**56/64**	j	**54/56**	k	**9/57/9**	l	**39/42/45**

DIVISION

Division sums can also often be simplified in this way.

7 **2)3 4 5 6** becomes 2)34 / 56 = **1728**,

Similarly **6)6 1 2** becomes 6)6 / 12 = **102**, (note the 0 here because the 12 takes up two places)

7)2 8 4 9 becomes 7)28 / 49 = **407**,

3) 2 4 4 5 3 becomes 3)24 / 45 / 3 = **8151**.

Practice D Divide the following mentally:

a 2)656	**b** 2)726	**c** 3)1821	**d** 6)1266
e 4)2048	**f** 4)2816	**g** 3)2139	**h** 2)2636

a	**3/28**	**b**	**36/3**	**c**	**6/07**	**d**	**2/11**
e	**5/12**	**f**	**7/04**	**g**	**7/13**	**h**	**13/18**

2.2 ALL FROM 9 AND THE LAST FROM 10

2.2a SUBTRACTION FROM A BASE

Applying the Vedic Sutra *All From 9 and the Last From 10* to a number gives another number.

8 If we apply the formula to **876** we get **124** because 8 and 7 are taken from 9 and 6 is taken from 10.

9 Similarly

3883	**64**	**98**	**6**	**10905**
↓	↓	↓	↓	↓
6117	**36**	**02**	**4**	**89095**

10 Applying the Sutra to **3450** or any number that ends in 0 we need to be a bit careful. If we take the last figure as 0 here, when we take it from 10 we get 10 which is a 2-figure number. To avoid this we **take 5 as the last figure**: we apply the Sutra to 345 and simply put the 0 on afterwards. So we get **6550**.

Similarly with 28160 we get **71840**,
and with 4073100 we get **5926900**.

Look again at the numbers in Examples 8, 9, 10 above.
If you add a number and its *All from 9...* number you should find that the total is always 10, 100, 1000, 10000 . . .
The total is always one of these unities, called **base numbers**.

This means that when we applied the Sutra to 876 we found how much it was below 1000.
Or, in other words, the Sutra gave the answer to the sum **1000 – 876**. The answer is **124**.

> The formula *All from 9 and the Last from 10* subtracts numbers from the next highest unity.

11 **1000 – 864 = 136** Just apply *All From 9 and the Last from 10* to 864,
1000 – 707 = 293,
10000 – 6523 = 3477,
100 – 76 = 24,
1000 – 580 = 420 Remember: apply the Sutra just to 58 here.

In every case the number is being subtracted from the next highest unity.

Practice E Subtract the following:

a 1000 – 481 **b** 1000 – 309 **c** 1000 – 892 **d** 1000 – 976

e 100 – 78 **f** 100 – 33 **g** 10000 – 8877 **h** 10000 – 9876

i 1000 – 808 **j** 1000 – 710 **k** 10000 – 6300

a	**519**	b	**691**	c	**108**	d	**24**
e	**22**	f	**67**	g	**1123**	h	**0124**
i	**192**	j	**290**	k	**3700**		

CALCULATIONS INVOLVING MONEY

We frequently need to make calculations of this type: when finding the change expected from a purchase for example.

12 Find **£10 – £4.56**

We simply apply the Sutra to 456 and give the answer: £10 – £4.56 = **£5.44**.

Practice F Subtract the following:

a £10 – £6.12 **b** £1 – £0.73 **c** £10 – £2.22 **d** £100 – £76.99

a	**£3.88**	b	**£0.27**	c	**£7.78**	d	**£23.01**

FIRST EXTENSION

In Exercise E you may have noticed that the number of zeros in the first number is the same as the number of figures in the number being subtracted.
E.g. 1000–481 has three zeros and 481 has three figures.

13 Suppose we had **1000 – 43**.
This has three zeros, but 43 is only a 2-figure number.
We can solve this by writing 1000 – 043 = **957**.
We put the extra zero in front of 43, and then apply the Sutra to 043.

14 **10000 – 58**.
Here we need to add two zeros: 10000 – 0058 = **9942**.

In the following exercise you will need to insert zeros, but you can do that mentally if you like.

Practice G Subtract the following:

a 1000 – 86 **b** 1000 – 29 **c** 1000 – 93 **d** 10000 – 678 **e** 10000 – 62

f 1000 – 8 **g** 10000 – 998 **h** £100 – £7.76 **i** £1 – £0.08

---

a	**914**	b	**971**	c	**907**	d	**9322**	e	**9938**
f	**992**	g	**9002**	h	**£92.24**	i	**£0.92**		

SECOND EXTENSION

15 Now consider **600 – 77**.
We have 600 instead of 100.
In fact the 77 will come off one of those six hundreds, so that 500 will be left.
So 600 – 77 = **523**.
The 6 is reduced by one to 5, and the Sutra is applied to 77 to give 23.

This reduction of one illustrates the Sutra *By One Less than the One Before*.

16 **5000 – 123 = 4877**. The 5 is reduced by one to 4,
and the Sutra converts 123 to 877.

17 Subtractions like this frequently arise when buying things.
If we offer a **£20** note when paying for goods costing **£3.46** the change we get should be **£16.54**.

We could think of the sum as being the same as 2000 – 346 (in pence), so the 2 is reduced to 1, and *All from 9...* is applied to 346.

18 **8000 – 3222**.

Considering the thousands there will be 4 left in the thousands column because we are taking over 3 thousand away.
All from 9... is then applied to the 222 to give 778.
So 8000 – 3222 = **4778**.

Practice H Subtract the following:

a 600 – 88 **b** 400 – 83 **c** 900 – 73 **d** 500 – 31

e 3000 – 831 **f** 8000 – 3504 **g** 2000 – 979 **h** £20 – £7.44

i £50 – £7.33

a 512	b 317	c 827	d 469
e 2169	f 4496	g 1021	h £12.56
i £42.67			

COMBINING THE FIRST AND SECOND EXTENSIONS

19 **6000 – 32**.
You will see here that we have a 2-figure number to subtract from 6000 which has three zeros. We will need to apply both of our extensions together.
The sum can be written 6000 – 032.
Then 6000 – 032 = **5968**.

The 6 is reduced to 5, and the Sutra converts 032 to 968.

20 **30000 – 63** = 30000 – 0063 = **29937**.
the 3 becomes 2, and 0063 becomes 9937.

Practice I Subtract the following:

a 5000 – 74	**b** 8000 – 58	**c** 3000 – 43	**d** 7000 – 81
e 6000 – 94	**f** 4000 – 19	**g** 80000 – 345	**h** 30000 – 276

a 4926	b 7942	c 2957	d 6919
e 5906	f 3981	g 79655	h 29724

The final exercise is a mixture of all the types:

Practice J Subtract:

a 100 – 34	**b** 1000 – 474	**c** 5000 – 542	**d** 800 – 72
e 1000 – 33	**f** 5000 – 84	**g** 700 – 58	**h** 9000 – 186
i 10000 – 4321	**j** 200 – 94	**k** 10000 – 358	**l** 400 – 81
m 7000 – 88	**n** 900 – 17	**o** 30000 – 63	**p** 90000 – 899

a 66	b 526	c 4458	d 728
e 967	f 4916	g 642	h 8814
i 5679	j 106	k 9642	l 319
m 6912	n 883	o 29937	p 89101

This leads to a general method of subtraction (See Section 2.2c).

2.2b BAR NUMBERS

The number **19** is very close to **20**.
And it can therefore be conveniently written in a different way: as $\mathbf{2\bar{1}}$.

21 $2\bar{1}$ means 20 – 1 = 19, the minus is put on top of the 1.

Similarly $3\bar{1}$ means 30 – 1 or 29.
And $4\bar{2}$ means 38.

We pronounce $4\bar{2}$ as "four, bar two" because the 2 has a bar on top.

This is rather like telling the time: we often say 'quarter to seven' or 'ten to seven' instead of 6:45 or 6:50 and so on.

These bar numbers can be very useful: they mean we can remove all the digits in a number which are over 5 and they give us a choice about different ways of representing a number.

22 $\mathbf{86\bar{1}}$ = 859, because $6\bar{1}$ = 59 (the 8 is unchanged),

$\mathbf{127\bar{3}}$ = 1267, because $7\bar{3}$ = 67.

Practice K Convert the following numbers:

a $6\bar{1}$	**b** $8\bar{2}$	**c** $3\bar{3}$	**d** $9\bar{1}$	**e** $46\bar{2}$	**f** $85\bar{1}$	**g** $774\bar{1}$	**h** $999\bar{1}$
i $1\bar{2}$	**j** $11\bar{1}$	**k** $12\bar{3}$	**l** $3\bar{4}0$				

a **59**	b **78**	c **27**	d **89**	e **458**	f **849**	g **7739**	h **9989**
i **8**	j **109**	k **117**	l **260**				

We may also need to **put numbers into bar form**.

23 **79 =** $\mathbf{8\bar{1}}$ because 79 is 1 less than 80,
239 = $\mathbf{24\bar{1}}$ because 39 = $4\bar{1}$,
508 = $\mathbf{51\bar{2}}$, 08 becomes $1\bar{2}$.

Practice L Put the following into bar form:

a 49	**b** 58	**c** 77	**d** 88	**e** 69	**f** 36	**g** 17
h 359	**i** 848	**j** 7719	**k** 328	**l** 33339	**m** 609	**n** 708

a $\mathbf{5\bar{1}}$	b $\mathbf{6\bar{2}}$	c $\mathbf{8\bar{3}}$	d $\mathbf{9\bar{2}}$	e $\mathbf{7\bar{1}}$	f $\mathbf{4\bar{4}}$	g $\mathbf{2\bar{3}}$
h $\mathbf{36\bar{1}}$	i $\mathbf{85\bar{2}}$	j $\mathbf{772\bar{1}}$	k $\mathbf{33\bar{2}}$	l $\mathbf{3334\bar{1}}$	m $\mathbf{61\bar{1}}$	n $\mathbf{71\bar{2}}$

24 How would you remove the bar number in $\mathbf{5\bar{1}3}$?
The best way is to split the number into two parts: $5\bar{1}/3$
Since $5\bar{1} = 49$, the answer is **493**.

> If a number has a bar number in it, split the number after the bar.

25 $\mathbf{7\bar{3}1} = 7\bar{3}/1 = \mathbf{671}$,

$\mathbf{52\bar{4}2} = 52\bar{4}/2 = \mathbf{5162}$,

$\mathbf{3\bar{2}15} = 3\bar{2}/15 = \mathbf{2815}$ since $3\bar{2} = 28$,

$\mathbf{5\bar{1}3\bar{2}} = 5\bar{1}/3\bar{2} = \mathbf{4928}$, as $5\bar{1} = 49$ and $3\bar{2} = 28$,

$\mathbf{3\bar{1}3\bar{2}3\bar{3}} = 3\bar{1}/3\bar{2}/3\bar{3} = \mathbf{292827}$.

Practice M Remove the bar numbers:

a $6\bar{1}4$	**b** $4\bar{2}3$	**c** $5\bar{2}5$	**d** $3\bar{1}7$	**e** $45\bar{2}3$	**f** $23\bar{4}5$	**g** $2\bar{2}2$
h $333\bar{2}3$	**i** $5\bar{1}32$	**j** $2\bar{3}55$	**k** $5\bar{4}4321$	**l** $4\bar{1}3\bar{1}$	**m** $6\bar{2}7\bar{3}$	**n** $2\bar{1}1$
o $41\bar{3}1$	**p** $52\bar{3}3$	**q** $7\bar{1}52$	**r** $1\bar{3}15\bar{1}$	**s** $9\bar{2}8\bar{3}$	**t** $1\bar{3}1$	**u** $13\bar{1}51$

a **594**	b **383**	c **485**	d **297**	e **4483**	f **2265**	g **182**
h **33283**	i **4932**	j **1755**	k **464321**	l **3929**	m **5867**	n **191**
o **4071**	p **5173**	q **6952**	r **7149**	s **8877**	t **71**	u **12951**

So far we have only had a bar on a single figure.
But we could have two or more bar numbers together.

26 Remove the bar numbers in $\mathbf{5\overline{33}}$.

The 5 means 500, and $\overline{33}$ means 33 is to be subtracted.
So $5\overline{33}$ means 500 – 33, and we have just seen sums of this type.

500 – 33 = **467** because the 33 comes off one of the hundreds, so the 5 is reduced to 4.
And applying *All from 9 and the Last from 10* to 33 gives 67.

27 Similarly $\mathbf{7\overline{14}} = \mathbf{686}$. The 7 reduces to 6 and the Sutra converts 14 to 86.

$\mathbf{26\overline{21}} = \mathbf{2579}$. 26 reduces to 25.

$\mathbf{7\overline{02}} = \mathbf{698}$. The Sutra converts 02 to 98.

$\mathbf{50\overline{3}} = \mathbf{497}$. 50 is reduced to 49 (alternatively, write $50\overline{3}$ as $5\overline{03}$: see previous example).

$\mathbf{4\overline{20}} = 4\overline{2}0 = \mathbf{380}$.

28 $\mathbf{4\overline{23}1}$ Here we can split the number after the bar: $4\overline{23}/1$.

$4\overline{23}$ changes to 377, and we just put the 1 on the end: $4\overline{23}1 = \mathbf{3771}$.

29 Similarly $\mathbf{5\overline{12}4} = 5\overline{12}/4 = \mathbf{4884}$,

$\mathbf{3\overline{11}33} = 3\overline{11}/33 = \mathbf{28933}$,

$\mathbf{5\overline{123}} = \mathbf{4877}$,

$\mathbf{3\overline{1}4\overline{31}} = 3\overline{1}/4\overline{31} = \mathbf{29369}$.

Practice N Remove the bar numbers:

a $6\overline{12}$	**b** $7\overline{33}$	**c** $2\overline{31}$	**d** $5\overline{11}$	**e** $9\overline{04}$	**f** $70\overline{6}$
g $55\overline{23}$	**h** $72\overline{41}$	**i** $333\overline{22}$	**j** $6\overline{21}4$	**k** $5\overline{31}22$	**l** $33\overline{22}44$
m $7\overline{333}$	**n** $6\overline{123}$	**o** $5\overline{104}$	**p** $44\overline{112}$	**q** $74\overline{031}$	**r** $7\overline{103}1$
s $6\overline{3322}$	**t** $3\overline{1}10\overline{2}$	**u** $4\overline{222}3$	**v** $3\overline{11}4\overline{1}$	**w** $3\overline{2}1\overline{22}$	**x** $310\overline{23}$

a **588**	b **667**	c **169**	d **489**	e **896**	f **694**
g **5477**	h **7159**	i **33278**	j **5794**	k **46922**	l **327844**
m **6667**	n **5877**	o **4896**	p **43888**	q **73969**	r **68971**
s **56678**	t **29098**	u **37783**	v **28939**	w **28078**	x **30977**

"The Sutras are very short; but, once one understands them and the modus operandi inculcated therein for their practical application, the whole thing becomes a sort of children's play and ceases to be a 'problem'."
From "Vedic Mathematics", Page 13.

ADVANTAGES OF BAR NUMBERS

Bar numbers are an ingenious device which we will be using in later work. Their main advantages are:

1. It gives us flexibility: we use the vinculum when it suits us.
2. Large numbers, like 6, 7, 8, 9 can be avoided.
3. Figures tend to cancel each other, or can be made to cancel.
4. 0 and 1 occur twice as frequently as they otherwise would.

The bar on top of a number is called a vinculum.

2.2c GENERAL SUBTRACTION

Pupils sometimes subtract in each column in a subtraction sum regardless of whether the top is greater than the bottom or not.
This method can however be used to give the correct answer.

30

$$\begin{array}{r} 4\,4\,4 \\ \underline{2\,8\,6} \; - \\ \underline{} \end{array}$$

Subtracting in each column we get $4-2=2$, $4-8=-4=\bar{4}$, $4-6=-2=\bar{2}$.
Since these negative answers can be written with a bar on top we can write:

$$\begin{array}{r} 4\,4\,4 \\ \underline{2\,8\,6} \; - \\ \underline{2\,\bar{4}\,\bar{2}} \end{array}$$

and $2\overline{42}$ is easily converted into **158**.

31 Similarly

$$\begin{array}{r} 6\,7\,6\,7 \\ \underline{1\,9\,0\,8} \; - \\ \underline{5\,\bar{2}\,6\,\bar{1}} \end{array} = \mathbf{4859}.$$

All subtraction sums can be dealt with by this method. We simply subtract each number from the number above it, putting a bar on the answer if the top is less than the bottom.
We then remove the bar numbers as shown before.

✏ **Practice O** Subtract using bar numbers:

a $\begin{array}{r} 5\,4\,3 \\ \underline{1\,6\,8} \; - \\ \underline{} \end{array}$ **b** $\begin{array}{r} 5\,6\,7 \\ \underline{2\,7\,9} \; - \\ \underline{} \end{array}$ **c** $\begin{array}{r} 8\,0\,4 \\ \underline{3\,8\,8} \; - \\ \underline{} \end{array}$ **d** $\begin{array}{r} 7\,3\,7 \\ \underline{5\,5\,8} \; - \\ \underline{} \end{array}$

e
```
 6413
 1878 −
 ____
```

f
```
 8024
 5339 −
 ____
```

g
```
 6543
 2881 −
 ____
```

h
```
 7103
 3991 −
 ____
```

i
```
 45654
 27986 −
 _____
```

j
```
 633
  88 −
 ___
```

k
```
 822
 577 −
 ___
```

l
```
 555
 275 −
 ___
```

---

a	**375**	**b**	**288**	**c**	**416**	**d**	**179**
e	**4535**	**f**	**2685**	**g**	**3662**	**h**	**3112**
i	**17668**	**j**	**545**	**k**	**245**	**l**	**280**

LESSON 3
SPECIAL METHODS

SUMMARY

3.1 ***Proportionately*** – various simple devices for multiplying and dividing using proportion.
3.2 ***All from 9 and the Last from 10*****: Multiplication** – multiplying and squaring numbers near a base.

If there is an easy way to do a particular sum, rather than using the general method, we call it a special method. For example to multiply a number by 10 we do not use 'long multiplication'. In the Vedic system there are many special methods, which adds to the fun: the general method is always there but there is often a quick way if you can spot it (see the note at the end of this chapter).

3.1 PROPORTIONATELY

The Vedic Sutra *Anurupyena*, which means *Proportionately*, is the first sub-Sutra of Vedic Mathematics.

3.1a DOUBLING AND HALVING

Doubling and halving numbers is very easy to do and can help us in many ways.

In the following exercise just write down the answers to the sums. Note that in halving a number like 56 we can find half of 50 and half of 6 and add them up to get 25 + 3 = 28.

Practice A

Double the following numbers:

a 43	**b** 1234	**c** 17	**d** 71	**e** 77	**f** 707
g 95	**h** 59	**i** 38	**j** 55	**k** 610	**l** 383

Halve the following numbers:

m 64	**n** 820	**o** 36	**p** 52	**q** 94	**r** 126
s 234	**t** 416	**u** 38	**v** 156	**w** 456	**x** 57

a 86	**b 2468**	**c 34**	**d 142**	**e 154**	**f 1414**
g 190	**h 118**	**i 76**	**j 110**	**k 1220**	**l 766**
m 32	**n 410**	**o 18**	**p 26**	**q 47**	**r 63**
s 117	**t 208**	**u 19**	**v 78**	**w 228**	**x 28½**

1 Find **35 × 4**.

We simply double 35 to 70, then double 70 to 140.
So **35 × 4 = 140**.

2 Find **26 × 8**.

Doubling 26 gives 52, doubling 52 gives 104, doubling 104 gives 208.
So **26 × 8 = 208**.

Practice B Multiply the following, mentally:

a 53 × 4 **b** 28 × 4 **c** 33 × 4 **d** 61 × 4 **e** 18 × 4 **f** 81 × 4

g 4½ × 4 **h** 20½ × 4 **i** 16¼ × 4 **j** 17 × 8 **k** 22 × 8 **l** 45 × 8

a 212	**b 112**	**c 132**	**d 244**	**e 72**	**f 324**
g 18	**h 82**	**i 65**	**j 136**	**k 176**	**l 360**

3 Divide **72 by 4**.

We halve 72 twice: half of 72 is 36, half of 36 is 18. So **72 ÷ 4 = 18**.

4 Divide **104 by 8**.

Here we halve three times: half of 104 is 52, half of 52 is 26, half of 26 is 13.
So **104 ÷ 8 = 13**.

Practice C

Divide by 4: **a** 56 **b** 68 **c** 84 **d** 180 **e** 116 **f** 92 **g** 34

Divide by 8: **h** 120 **i** 440 **j** 248 **k** 216 **l** 44

a 14	**b 17**	**c 21**	**d 45**	**e 29**	**f 23**	**g 8½**
h 15	**i 55**	**j 31**	**k 27**	**l 5½**		

Another way of applying *Proportionately* is to combine doubling and halving.

5 Find **35 × 22**.

We double 35 and halve 22 and this gives us 70×11 which has the same answer as 35×22.
So 35 × 22 = 70 × 11 = **770**.

Practice D Multiply the following:

a 15 × 18 **b** 15 × 24 **c** 35 × 14 **d** 45 × 18 **e** 22 × 15 **f** 16 × 4$\frac{1}{2}$

g 24 × 3$\frac{1}{2}$ **h** £4.50 × 32 **i** 44 × 225 (halve and double, twice) **j** 32 × 325

a	270	b	360	c	490	d	810	e	330	f	72
g	84	h	£144	i	9900	j	10400				

3.1b EXTENDING THE MULTIPLICATION TABLES

You may have noticed another way of doing some of the sums in Practice B.

6 For example, for **18 × 4** you may have thought that since you know that 9 × 4 = 36, then 18×4 must be double this, which is **72**.

Similarly if you don't know **8 × 7** but you do know that 4 × 7 = 28, you can just double 28.
So **8 × 7 = 56**.

7 Find **6 × 14**.

Since we know that 6 × 7 = 42, it follows that **6 × 14 = 84**.

8 Find **14 × 18**.

Halving 14 and 18 gives 7 and 9, and since 7× 9 = 63 we double this twice.
We get 126 and 252, so **14 × 18 = 252**.

Practice E Multiply the following:

a 16 × 7 **b** 18 × 6 **c** 14 × 7 **d** 12 × 9 **e** 4 × 14 **f** 6 × 16

g 7 × 18 **h** 9 × 14 **i** 16 × 18 **j** 14 × 16 **k** 12 × 18 **l** 16 × 12

a	112	b	108	c	98	d	108	e	56	f	96
g	126	h	126	i	288	j	224	k	216	l	192

3.1c MULTIPLYING BY 5, 50, 25

The numbers 2 and 5 are closely associated because 2 × 5 = 10.

9 Find **44 × 5**.

We find half of 440, which is 220. So 44 × 5 = **220**.

10 Find **87 × 5**.

Half of 870 is 435. So 87 × 5 = **435**. Similarly 4.6 × 5 = half of 46 = 23.

11 Find **27 × 50**

We multiply 27 by 100, and halve the result. Half of 2700 is 1350.
So 27 × 50 = **1350**. Similarly 5.2 × 50 = half of 520 = 260.

12 Find **82 × 25**

25 is half of half of 100, so to multiply a number by 25 we multiply it by 100 and halve twice. So we find half of half of 8200, which is 2050.
82 × 25 = **2050**.

✎ **Practice F** Multiply the following:

a 68 × 5	**b** 42 × 5	**c** 36 × 5	**d** 56 × 5	**e** 61 × 5	**f** 426 × 5
g 803 × 5	**h** 2468 × 5	**i** 46 × 50	**j** 864 × 50	**k** 223 × 50	**l** 202 × 50
m 72 × 25	**n** 48 × 25	**o** 85 × 25	**p** 808 × 25		

---

a 340	**b 210**	**c 180**	**d 280**	**e 305**	**f 2130**
g 4015	**h 12340**	**i 2300**	**j 43200**	**k 11150**	**l 10100**
m 1800	**n 1200**	**o 2125**	**p 20200**		

"The Sutras are easy to understand, easy to apply and easy to remember; and the whole work can be truthfully summarised in one word "mental".
From "Vedic Mathematics", Page xvi.

3.2 *ALL FROM 9 AND THE LAST FROM 10*: MULTIPLICATION

Another useful application of the Sutra **Nikhilam** Navatascharaman Dasatah (*All from 9 and the Last from 10*) is in multiplying numbers which are close to a base number, like 10, 100, 1000 etc. We call this **Nikhilam multiplication** or Base multiplication.

3.2a NUMBERS JUST BELOW 100

13 Usually a sum like **88 × 98** is considered especially difficult because of the large figures, 8 and 9.

But since the numbers 88 and 98 are close to the base of 100 you may think that there ought to be a simple way to find such a product and in the Vedic system there is a wonderfully easy way.

88 – 12	88 is 12 below 100, so we put –12 next to it,
98 – 2	98 is 2 below 100 so we put –2 next to it.
86 / 24	We call the 12 and 2 **deficiencies** as the numbers 88 and 98 are deficient from the unity of 100 by 12 and 2.

The answer 8624 is in two parts: 86 and 24.
The 86 is found by taking one of the deficiencies from the other number: that is 88–2 = **86** or 98–12 = 86 (whichever you like),
and the 24 is simply the product of the deficiencies: 12 × 2 = **24**.
So **88 × 98 = 8624**. It could hardly be easier.

GEOMETRICAL PROOF

Let us look at a geometrical proof for this.
88×98 is the area of a rectangle 88 units by 98 units so we begin with a square of side 100:

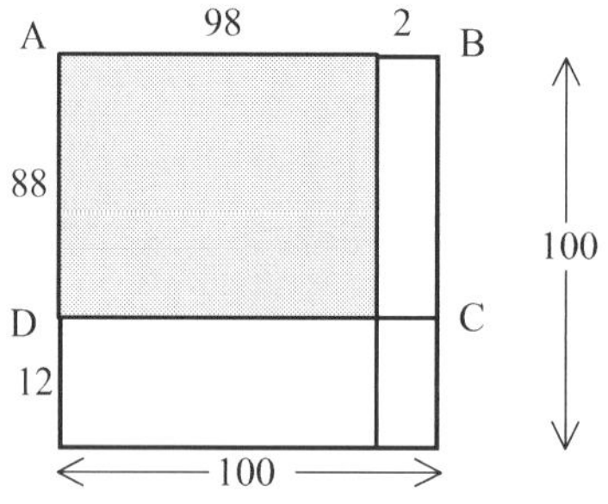

You can see the required area shaded in the diagram.
You can also see the deficiencies from 100: 12 and 2.

Now the area ABCD must be 8800 because the base is 100 and the height is 88.

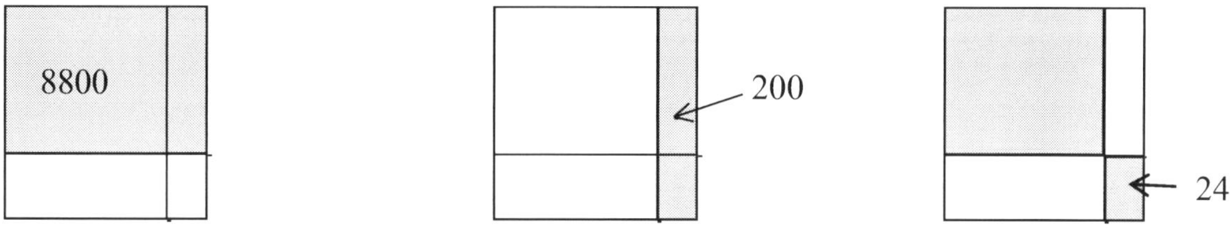

From this we subtract the strip on the right side, the area of which is 200: so 8800 – 200 = 8600.
This leaves the required area but we have also subtracted the area of the small rectangle shown shaded above on the bottom right. This must therefore be added back on and since its area is 12×2=24 we add 24 to 8600 to get **8624**.

ALGEBRAIC PROOF

(x – a)(x – b) = x(x – a – b) + ab
where x is the base number, 100 in this case, and a and b are the deficiencies, 12 and 2.

14 For **93 × 96** we get deficiencies of 7 and 4, so

93 – 07
96 – 4
89 / 28

The differences from 100 are 7 and 4,
93 – 4 = **89** or 96 – 7 = 89,
and 7 × 4 = **28**.

> In fact once we have got the deficiencies we apply the *Vertically and Crosswise* Sutra:
> **we cross-subtract** to get the left-hand part of the answer, and
> **we multiply vertically** in the right-hand column to get the right-hand part of the answer.

15 For **98 × 97**:

98 – 02
97 – 03
95 / 06

Note the zero inserted here: the numbers being multiplied are near to 100, so two digits are required on the right, as in the other examples.

16 For **89 × 89**:

89 – 11
89 – 11
78 / $_{1}$21 = **7921**

Here the numbers are each 11 below 100, and 11 × 11 = 121, a 3–figure number.
The hundreds digit of this is therefore carried over to the left.

Practice G Multiply the following:

a 94 × 94 **b** 97 × 89 **c** 87 × 99 **d** 87 × 98 **e** 87 × 95

f 95 × 95 **g** 79 × 96 **h** 98 × 96 **i** 92 × 99 **j** 88 × 88

k 97 × 56 **l** 97 × 63 **m** find a way of getting 92 × 196?

---

a	**88/36**	b	**86/33**	c	**86/13**	d	**85/26**	e	**82/65**
f	**90/25**	g	**75/84**	h	**94/08**	i	**91/08**	j	**77/44**
k	**54/32**	l	**61/11**	m	**180/32 (find 92×98 and double the answer)**				

The most efficient way to do these sums is to take one number and subtract the other number's deficiency from it. Then multiply the deficiencies together, mentally adjusting the first part of the answer if there is a carry figure.
This is so easy it is really just mental arithmetic.

Practice H Do these sums mentally, just write down the answer:

a	**b**	**c**	**d**	**e**	**f**	**g**
87	79	98	94	96	88	89
97	98	93	95	96	96	98
___	___	___	___	___	___	___

It is also worth practising these sums after only hearing them, rather than seeing them written down, so here are a few more.

h	**i**	**j**	**k**	**l**
88	87	97	96	95
98	97	97	92	85
___	___	___	___	___

---

a	**84/39**	b	**77/42**	c	**91/14**	d	**89/30**	e	**92/16**	f	**84/48**	g	**87/22**
h	**86/24**	i	**84/39**	j	**94/09**	k	**88/32**	l	**80/75**				

3.2b OTHER BASES

17 **568 × 998 = 566864.**

In this sum the numbers are close to 1000, and the deficiencies are 432 and 2.
The deficiency for 568 is found by applying the Nikhilam Sutra: *All from 9 and the Last from 10.*

568 – 432
998 – 2
566 / 864

The method here is just the same, but we allow 3 figures on the right as the base is now 1000.

The differences of the numbers from 1000 are 432 and 2.
Then cross-subtracting: 568 – 2 = **566**,
And vertically: 432 × 2 = **864**.

> The number of spaces needed on the right is the number of 0's in the base number.

18 **68778 × 99997**.

Even large numbers like this are easily and mentally multiplied by the same method.

68778 – 31222
99997 – 3
68775 / 93666

19 **7 × 8**.

In the Vedic system tables above 5×5 are not really essential:

Exactly the same method gives us 7 × 8 = 56.

7 – 3
8 – 2
5 / 6

✐ **Practice I** Multiply the following mentally:

a 667 × 998	**b** 768 × 997	**c** 989 × 998	**d** 885 × 997
e 883 × 998	**f** 8 × 6	**g** 891 × 989	**h** 8888 × 9996
i 6999 × 9997	**j** 90909 × 99994	**k** 78989 × 99997	**l** 9876 × 9989

---

a	**665/666**	b	**765/696**	c	**987/022**	d	**882/345**
e	**881/234**	f	**4/8**	g	**881/199**	h	**8884/4448**
i	**6996/9003**	j	**90903/54546**	k	**78986/63033**	l	**9865/1364**

3.2c NUMBERS ABOVE THE BASE

Suppose now that the numbers are not both below a base number as in all the previous examples, but above the base.

20 **103 × 104 = 10712**.

103 +03
104 + 4
107 / 12

This is even easier than the previous examples, but the method is just the same. The differences from the base are +3 and +4 since the numbers are now **above the base**.

103 + 4 = 107 or 104 + 3 = 107, and 4 × 3 = 12.

So now we **cross-add,** and multiply vertically.

21 **12 × 13 = 156**. (12+3=15, 2×3=6)

22 **1234 × 1003 = 1237702**. (1234+3=1237, 234×3=702)

23 **10021 × 10002 = 100230042**. (10021+2=10023, 0021×2=0042)

With a base of 10,000 here we need 4 figures on the right.

Practice J Multiply mentally:

a 133 × 103	**b** 107 × 108	**c** 171 × 101	**d** 102 × 104
e 123 × 102	**f** 14 × 12	**g** 18 × 13	**h** 1222 × 1003
i 1051 × 1007	**j** 15111 × 10003	**k** 125 × 105	**l** 10607 × 10008

---

a	**136/99**	b	**115/56**	c	**172/71**	d	**106/08**
e	**125/46**	f	**16/8**	g	**23/4**	h	**1225/666**
i	**1058/357**	j	**15115/5333**	k	**131/25**	l	**10615/4856**

Algebraic Proof: $(x + a)(x + b) = x(x + a + b) + ab$, where $x=10^n$.

3.2d ONE NUMBER ABOVE AND ONE BELOW THE BASE

24 Find **124 × 98**.

Here one number is over and the other is under 100:
The differences from 100 are +24 and –2.
Crosswise gives 122 (124–2 or 98+24).

$$\begin{array}{rl} 124 & +24 \\ \underline{98} & \underline{-\ 2} \\ 122\ / & \overline{48} \end{array} = \mathbf{12152}$$

So 122 is the left-hand part of the answer.
Then multiplying the differences we get –48, written $\overline{48}$ (since a plus times a minus gives a minus). This gives the answer as $122\overline{48}$.

To remove the negative portion of the answer we just take 48 from one of the hundreds in the hundreds column. This simply means reducing the hundreds column by 1 and applying *All From 9 and the Last From 10* to 48.
Thus 122 becomes 121 and $\overline{48}$ becomes 52.
So 124 × 98 = $122\overline{48}$ = **12152**.

25 **1003 × 987** = $990/\overline{039}$ = **989/961**.

Similarly, we first get 1003 – 13 = 990 or 987 + 3 = 990,
and +3 × –13 = $\overline{039}$ (three figures required here as the base is 1000).

Then 990 is reduced by 1 to 989, and applying the formula to 039 gives 961.
So these sums are just like the others except that we need to clear the minus part at the end.

26 **121 × 91** = $112/_{\bar{1}}\overline{89}$ = **110/11**.

Here we have a minus one to carry over to the left so that the 112 is reduced by 2 altogether.

Practice K

a 104 × 91 **b** 94 × 109 **c** 103 × 98 **d** 92 × 112

e 91 × 111 **f** 106 × 89 **g** 91 × 103 **h** 91 × 107

i 91 × 105 **j** 991 × 1005 **k** 987 × 1006 **l** 992 × 1111

a	**9464**	b	**10246**	c	**10094**	d	**10304**
e	**10101**	f	**9434**	g	**9373**	h	**9737**
i	**9555**	j	**995955**	k	**992922**	l	**1102112**

Algebraic Proof: $(x + a)(x - b) = x(x + a - b) - ab$, where $x=10^n$.

3.2e *PROPORTIONATELY*

The *Proportionately* formula considerably extends the range of this multiplication method.

27 **213 × 203 = 43239**.

$$\begin{array}{rr} & 213 + 13 \\ & \underline{203 + \ \ 3} \\ 2 \times & \underline{216\ /\ 39} = \mathbf{43239} \end{array}$$

We observe here that the numbers are not near any of the bases used before: 10, 100, 1000 etc.
But they are close to 200, with differences of 13 and 3 as shown above.

The usual procedure gives us 216/39 (213+3=216, 13×3=39).

Now since our base is 200 which is 100×2 we multiply **only the left-hand part** of the answer by 2 to get 43239.

Algebraic Proof: $(nx + a)(nx + b) = nx(x + a + b) + ab$, $x=10^n$.

28 **29 × 28 = 812**.

The base is 30 (3×10), and the deficiencies are –1 and –2.
Cross-subtracting gives 27,
then multiplying vertically on the right we get **2**,
and finally 3×27 = **81**.

$$\begin{array}{rr} & 29 - 1 \\ & \underline{28 - 2} \\ 3 \times & \underline{27\ /\ 2} \end{array} = \mathbf{812}$$

So these are just like the previous sums but with an extra multiplication (**of the left-hand side only**) at the end.

29 **33 × 34**.

In this example there is a carry figure:

$$\begin{array}{r} 33 + 3 \\ \underline{34 + 4} \\ \underline{3\times37\ /\ {}_{1}2 = 111/{}_{1}2} \end{array} = \mathbf{1122}$$

Note that since the right-hand side does not get multiplied by 3 we **multiply** the left-hand side by 3 **before adding the carried figure**.

30 **88 × 49** = ½(88×98) = ½(8624) = **4312**.

This example shows a different application of *Proportionately*.
In 88 × 49 the numbers are not both close to 100, but since twice 49 is 98 we can find 88 × 98 and halve the answer at the end.

Practice L Multiply mentally:

a 41 × 42	**b** 204 × 207	**c** 321 × 303	**d** 203 × 208
e 902 × 909	**f** 48 × 47	**g** 188 × 196	**h** 199 × 198
i 189 × 194	**j** 207 × 211	**k** 312 × 307	**l** 5003 × 5108
m 23 × 24	**n** 44 × 98	**o** 48 × 98	**p** 192 × 98

---

a	**172/2**	b	**422/28**	c	**972/63**	d	**422/24**
e	**8199/18**	f	**225/6**	g	**368/48**	h	**394/02**
i	**366/66**	j	**436/77**	k	**957/84**	l	**25555/324**
m	**55/2**	n	**43/12**	o	**47/04**	p	**188/16**

3.2f MULTIPLYING NUMBERS NEAR DIFFERENT BASES

31 **9998 × 94 = 9398/12**.

Here the numbers are close to different bases: 10,000 and 100, and the deficiencies are –2 and –6.

We write, or imagine, the sum set out as shown:

9998	–02
94	– 6
9398 /	12

It is important to line the numbers up as shown because the 6 is not subtracted from the 8, but from the 9 above the 4 in 94. That is, the second column from the left here.

Then multiply the deficiencies together: 2×6 = **12**.

Note that the number of figures in the right-hand part of the answer corresponds to the base of the lower number (94 is near 100, therefore there are 2 figures on the right).

You can see why this method works by looking at the sum 9998 × 9400, which is 100 times the sum done above:

9998	– 0002
9400	– 600
9398	/ 1200

Now we can see that since 9998 × 9400 = 93981200,
then 9998 × 94 = 939812.

This also shows why the 6 above is subtracted in the second column from the left.

32 **10007 × 1003 = 10037021**.

Lining the numbers up:

10007	+ 007
1003	+ 3
10037 /	021

we see that we need three figures on the right and that the surplus, 3, is added in the 4th column, giving 10037.

Practice M Find:

a 97 × 993 **b** 92 × 989 **c** 9988 × 98 **d** 9996 × 988

e 103 × 1015 **f** 106 × 1012 **g** 10034 × 102 **h** 1122 × 104

---

a **963/21** b **909/88** c **9788/24** d **9876/048**
e **1045/45** f **1072/72** g **10234/68** h **1166/88**

Algebraic Proof: $(x + a)(y + b) = (x + a)y + bx + ab$, where $x=10^m$, $y=10^n$.

Another application of this type of multiplication is in multiplying three or more numbers simultaneously which are close to a base (se Reference 3).

3.2g SQUARING NUMBERS NEAR A BASE

This is especially easy and is for squaring numbers which are near 10, 100, 1000 etc.

You will recall that squaring means that a number is multiplied by itself (like 96×96).

This method is described by the sub-formula *Reduce (or increase) by the Deficiency and also set up the square.*

33 **96^2 = 92/16**

96 is 4 below 100, so we reduce 96 by 4, which gives us the first part of the answer, **92**.
The last part is just 4^2 =**16**, as the formula says.

34 **1006^2 = 1012/036**

Here 1006 is increased by 6 to 1012, and $6^2 = 36$: but with a base of 1000 we need 3 figures on the right, so we put 036.

35 **304^2** = 3×308/16 = **92416**

This is the same but because our base is 300 the left-hand part of the answer is multiplied by 3.

✎ **Practice N** Square the following:

a 94	**b** 103	**c** 108	**d** 1012	**e** 98	**f** 88
g 91	**h** 10006	**i** 988	**j** 997	**k** 9999	**l** 9989
m 111	**n** 13	**o** 212	**p** 206	**q** 302	**r** 601
s 21	**t** 4012	**u** 511	**v** 987		

w What number, when squared, gives 9409?

x What number, when squared, gives 11449?

a	**8836**	**b**	**10609**	**c**	**11664**	**d**	**1024144**	**e**	**9604**	**f**	**7744**
g	**8281**	**h**	**100120036**	**i**	**976144**	**j**	**994009**	**k**	**99980001**	**l**	**99780121**
m	**12321**	**n**	**169**	**o**	**44944**	**p**	**42436**	**q**	**91204**	**r**	**361201**
s	**441**	**t**	**16096144**	**u**	**261121**	**v**	**974169**	**w**	**97**	**x**	**107**

Algebraic Proof: $(x + a)^2 = (x + 2a) + a^2,$
$(nx + a)^2 = n(nx + a)x + a^2.$

3.3 MENTAL CALCULATIONS

The Vedic techniques are so easy that the system of Vedic Mathematics is really a system of mental mathematics. This has a number of further advantages as pupils seem to make faster progress and enjoy mathematics more when they are permitted to do the calculation in their head. After all, the objects of mathematics are mental ones, and writing down requires a combination of mental and physical actions, so that the child's attention is alternating between the mental and physical realms. This alternation is an important ability to develop but working only with mental objects also has many advantages.

Mental mathematics leads to greater creativity and the pupils understand the objects of mathematics and their relationships better. They begin to experiment (especially if they are encouraged to do so) and become more flexible. Memory and confidence are also improved through mental mathematics.

SPECIAL METHODS

The special methods play a large part in encouraging mental mathematics. Everyone likes a short cut, whether it is a quick way to get from one place to another or an easy way of doing a particular calculation. Life is full of special methods: to tackle all similar situations in the same way is not the way most people like to function. Every mathematical calculation invites its own unique method of solution and we should encourage children to look at the special properties of each problem in order to understand it best and decide on the best way forward. This is surely the intelligent way to do mathematics.

LESSON 4
BY ONE MORE THAN THE ONE BEFORE

SUMMARY

4.1 SPECIAL MULTIPLICATIONS

The Vedic formulas give us some very fast methods of calculation.
In particular the Sutra *Ekadhikena Purvena*: *By One More than the One Before* gives us some special multiplication and division devices which are extremely efficient.

4.1a SQUARING NUMBERS THAT END IN 5

The formula *By One More Than the One Before* provides a beautifully simple way of squaring numbers that end in 5.

1 In the case of $\mathbf{75^2}$, we simply multiply the 7 (the number before the 5) by the next number up, 8. This gives us 56 as the first part of the answer, and the last part is simply 25 (5^2).

So $\mathbf{75^2 = 56/25}$ where $56=7\times8$, $25=5^2$.

2 Similarly $\mathbf{65^2 = 4225}$ $42=6\times7$, $25=5^2$.

3 And $\mathbf{25^2 = 625}$ where $6=2\times3$.

4 Also since $4\frac{1}{2}= 4.5$, the same method applies to squaring numbers ending in $\frac{1}{2}$.
So $\mathbf{4\frac{1}{2}^2 = 20\frac{1}{4}}$, where $20 = 4\times5$ and $\frac{1}{4} = \frac{1}{2}^2$.

The method can be applied to numbers of any size:

5 $\mathbf{305^2 = 93025}$ where 930 = 30×31.

Even for squaring large numbers like, say, 635, it is still easier to multiply 63 by 64 and put 25 on the end than to multiply 635 by 635.

Practice A Square the following numbers:

a 55	**b** 15	**c** 8½	**d** 95	**e** 105
f 195	**g** 155	**h** 245	**i** 35	**j** 20½
k 8005	**l** 350	**m** What number, when squared gives 2025?		

---

a 3025	b 225	c 72¼	d 9025	e 11025
f 38025	g 24025	h 60025	i 1225	j 420¼
k 64080025	l 122500	m 45		

Algebraic Proof: $(ax + 5)^2 = a(a + 1)x^2 + 25$, where x = 10. See also Page 69.

4.1b A VARIATION

6 Suppose we want to find **43 × 47** in which both numbers begin with the same figure, 4, and the last figures (3 and 7) add up to 10.

The method is just the same as in the previous section:
multiply 4 by the number *One More*, 4 × 5 = 20.

Then simply multiply the last figures together: 3 × 7 = 21.
So **43 × 47 = 2021** where 20 = 4×5, 21 = 3×7.

7 Similarly **62 × 68 = 4216** where 42 = 6×7, 16 = 2×8.

8 **204 × 206**: both numbers start with 20, and 4 + 6 = 10, so the method applies here:

204 × 206 = **42024**, (420 = 20×21, 24 = 4×6).

9 **93 × 39** may not look like it comes under this particular type of sum, but remembering the *Proportionately* formula we notice that 93 = 3×31, and 31 × 39 does come under this type:

31 × 39 = 1209 (we put 09 as we need double figures here),

so **93 × 39 = 3627** (multiply 1209 by 3).

The thing to notice is that the 39 needs a 31 for the method to work here: and then we spot that 93 is 3×31.

10 Finally, consider **397 × 303**.

Only the 3 at the beginning of each number is the same, but the rest of the numbers (97 and 03) add up to 100.
So again the method applies, but this time we must expect to have four figures on the right-hand side:
3/97 × 3/03 = 120291 where 12 = 3×4, 0291 = 97×3.

Practice B Multiply the following:

a 73 × 77	**b** 58 × 52	**c** 81 × 89	**d** 104 × 106
e 297 × 293	**f** 303 × 307	**g** 64 × 38	**h** 88 × 46
i 33 × 74	**j** 66 × 28	**k** 36 × 78	**l** 46 × 54
m 298 × 202	**n** 391 × 309	**o** 795 × 705	**p** 401 × 499

a	**5621**	b	**3016**	c	**7209**	d	**11024**
e	**87021**	f	**93021**	g	**2432**	h	**4048**
i	**2442**	j	**1848**	k	**2808**	l	**2484**
m	**60196**	n	**120819**	o	**560475**	p	**200099**

Algebraic Proof: $(ax + b)(ax + 10-b) = a(a + 1)x^2 + b(10 - b)$. See also Page 69.

4.2 MULTIPLICATION SUMMARY

Here we summarise most of the methods of multiplication and squaring encountered so far.

1. We can multiply from left to right using *On the Flag*. E.g. 456×3.

2 Multiplying by 4, 8 etc. we can just double twice, 3 times etc. E.g. 37×4=148.

3 We can use doubling to extend the multiplication tables. E.g. 14×8=112.

4 We can use halving to multiply by 5, 50 and 25. E.g. 36×5=180.

5 We can use *All from 9 and the Last from 10* for multiplying numbers near a base, or near different bases.

E.g. 98×88=8624, 103×104=10712, 1123×997=1119631, 203×204=41412, 9988×98.

6 The same Sutra can be used for squaring numbers near a base. E.g. 97^2, 1006^2, 203^2.

7 We can square numbers ending in 5, using *By One More than the One Before*. E.g. 35^2.

8 With the same Sutra we can multiply numbers that have the same first digit and whose last figures add up to 10. E.g. 72×78.

The following exercise contains a mixture of all these different types of multiplication:

✎ **Practice C**

a 654 × 3	**b** 86 × 98	**c** 97 × 92	**d** 73 × 4	**e** 7 × 22
f 16 × 24	**g** 798 × 997	**h** 8899 × 9993	**i** 86 × 5	**j** 84 × 25
k 103 × 109	**l** 123 × 96	**m** 203 × 209	**n** 188 × 197	**o** 85^2
p 73 × 77	**q** 32 × 33	**r** 2004 × 2017	**s** 9997 × 98	**t** 1023 × 102
u 84 × 86	**v** 28 × 54	**w** 303 × 307	**x** 93^2	**y** 1011^2
z 403^2				

a **1,962**	b **8,428**	c **8,924**	d **292**	e **154**
f **384**	g **795,606**	h **88,927,707**	i **430**	j **2100**
k **11,227**	l **11,808**	m **42,427**	n **37,036**	o **7,225**
p **5,621**	q **1,056**	r **4,042,068**	s **979,706**	t **104,346**
u **7,224**	v **1,512**	w **93,021**	x **8,649**	y **1,022,121**
z **162,409**				

The methods covered so far are special methods which apply in special types of sum. In Section 6.2 we look at the general multiplication method by which any numbers can be multiplied in one line, from right to left or from left to right.

"all that the student has to do is to look for certain characteristics, spot them out, identify the particular type and apply the formula which is applicable thereto."
From "Vedic Mathematics", Page 106.

4.3 RECURRING DECIMALS

A fraction can be converted to a decimal by dividing the numerator by the denominator. This can be quite time consuming if the denominator is not small, but the Vedic system gives us some really easy and beautiful methods which are also extremely fast.

4.3a DENOMINATOR ENDING IN 9

11 Convert the fraction $\frac{1}{19}$ to its decimal form.

Dividing by 19 is normally not too easy but here we can use the formula *By One More than the One Before.*
This means one more than the number before the 9 in this fraction.
Since there is a 1 before the 9, one more than this is **2**.
So for this fraction the Sutra says ***By 2***.

We call **2** the **Ekadhika** (the number *one more*) and we keep dividing by this **2** (rather than dividing by 19). 'Ekadhika' is pronounced with two long syllables followed by two short syllables, the long syllables being twice the length of the short ones.

So we start with 0 and a decimal point. $\frac{1}{19} = 0.$

Then dividing **2** into 1 (the numerator) goes 0 remainder 1.

Note carefully that we put the remainder **before** the answer, 0. $\frac{1}{19} = 0._{1}0$

We now have 10 in front of us ($_{1}0$) and we divide this by **2**: $\frac{1}{19} = 0._{1}05$

We then divide this 5 by **2**.

This gives 2 remainder 1 so we now have: $\frac{1}{19} = 0._{1}05_{1}2$

Again we put the remainder, 1, before the 2.

> At every step we divide the last answer figure by the Ekadhika, **2**, and put the answer down as the next answer figure.
> Any remainder is put before that answer figure.

Then we divide 12 by **2** and put down 6.

$$\frac{1}{19} = \mathbf{0._{1}\,\dot{0}5_{1}263_{1}1_{1}5_{1}7_{1}89_{1}47_{1}3_{1}6842\dot{1}}$$

We find that after 18 figures the answer figures are starting to repeat themselves (we get 10 to divide and then 5, which is how we started the sum).
This means that those 18 figures will repeat themselves forever.

To show that the 18 figures repeat endlessly we put a dot over the first and last figures.
These dots indicate the first and last numbers of the block which repeats indefinitely.

The answer is $\frac{1}{19} = \mathbf{0.\dot{0}5263157894736842\dot{1}}$, the remainder figures are not part of the answer.

Dividing by 2 like this is very much easier than dividing by 19 of course!

12 Convert $\frac{11}{19}$ to a recurring decimal.

The Ekadhika is still **2** because we still have 19 in the denominator.
But we begin by dividing **2** into 11 (the numerator) this gives 5 remainder 1:

$$\frac{11}{19} = 0._{1}5$$

Next we divide **2** into 15: $\frac{11}{19} = 0._{1}5_{1}7$

Continuing the division until it starts to repeat we get:

$$\frac{11}{19} = \mathbf{0.\dot{5}7894736842105263\dot{1}}$$

There are two important things which you may have noticed as you worked out $\frac{11}{19}$.
It also has 18 figures recurring like $\frac{1}{19}$.
In fact the figures are the same as $\frac{1}{19}$ but they just start in a different place.

Block recurrers like this can be plotted on the 9-point circle, each denominator having its own individual pattern or patterns.

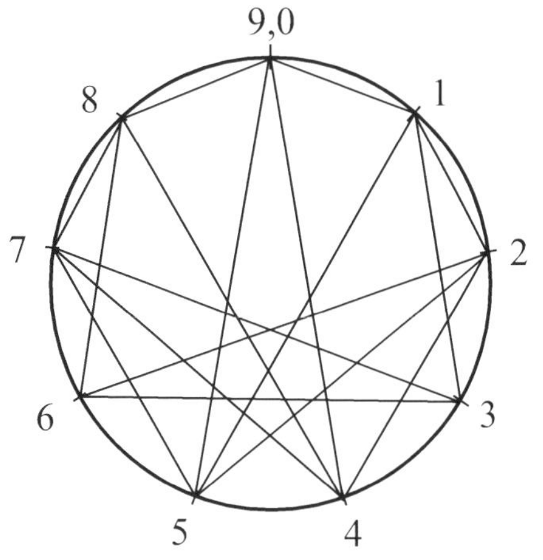

To plot the recurring decimal for $\frac{1}{19}$ we find the first figure in the block on the 9-point circle, this is 0. We then join this to the next figure in the sequence, 5, and then to 2 and so on.

Remember when you get to the 1 at the end you have one more line to draw: because the decimal repeats in a cycle of 18 figures, the 1 at the end is followed by the first figure which is 0. You should have a diagram which is symmetrical about a vertical axis.

For drawing these patterns you will find some 9-point circles at the end of this book.

There is no need to draw the pattern for $\frac{11}{19}$ because it is really the same cycle (just starting at a different place) and so will look identical, when it is finished, to the pattern for $\frac{1}{19}$.

13 Note that the decimal for $\frac{11}{19}$ can be obtained from the decimal for $\frac{1}{19}$ because $\frac{11}{19}=\frac{10}{19}+\frac{1}{19}$ and $\frac{10}{19}$, being ten times $\frac{1}{19}$ will be $\frac{10}{19}=0.\dot{5}2631578947368421\dot{0}$. So we can just add these two decimals, which can be done by adding pairs of digits in the decimal for $\frac{1}{19}$: 0+5=**5**, 5+2=**7**, 2+6=**8** . . . (rather like multiplying by 11 - see Section 8.4).

PROOF

In dividing 19 into an even figure followed by a zero:

$$19\overline{)20}\;\;1r1 \qquad 19\overline{)40}\;\;2r2 \qquad 19\overline{)60}\;\;3r3 \qquad \text{etc.}$$

we find the quotient is equal to the remainder, and half the even number.
This is because 19 = $2\bar{1}$. And since in decimal division the remainder with zero appended becomes the next dividend, we can divide into the quotient instead. If the dividend is an odd number we carry one to the right, as 10.

An alternative explanation is supplied by the well-known result: $\frac{1}{x-1}=\frac{1}{x}+\frac{1}{x^2}+\frac{1}{x^3}+\ldots$

which, for x = 20 becomes: $\frac{1}{19}=\frac{1}{20}+\frac{1}{20^2}+\frac{1}{20^3}+\ldots$ Here each term is obtained from the previous term by dividing it by 2 and putting it one place to the right (dividing by 10).

14 Convert $\mathbf{\frac{17}{29}}$ to decimal form.

Here the number before the 9 is 2 so *One More than the One Before* means one more than 2, which is **3**. So **the Ekadhika is now 3**.
This means we start by dividing 17 (the numerator) by **3**, and keep on dividing by **3**: **3** into 17 goes 5 remainder 2.

$$\frac{17}{29}=0._25$$

Then **3** into 25 goes 8 remainder 1; and **3** into 18 goes 6 and so on.

The full recurring decimal for $\frac{17}{29}$ is $0.\dot{5}86206896551724137931034482\dot{7}$.
There are 28 figures here.

❋ Plot the pattern for $\frac{17}{29}$.

\# You may notice that the number of figures in the recurring decimals seen so far is one less than the denominator: (N=D–1). This is often the case, but not always: $\frac{9}{39}$ has six recurring figures, not 38.
But the number of figures cannot exceed D–1. In dividing 1 by 19, for example, there can be only 18 possible different remainders and so there cannot be more than 18 figures in the recurring decimal.

4.3b A SHORT CUT

You have now worked out three recurring decimals: $\frac{1}{19}$, $\frac{11}{19}$, $\frac{17}{29}$ using the Ekadhika method. The first two have 18 recurring figures and the third has 28.

If you write out the three recurring decimals but with $\frac{1}{19}$ and $\frac{11}{19}$ in two rows of 9 figures and $\frac{17}{29}$ with two rows of 14 figures you may notice something:

$$\frac{1}{19} = 0.{}_1\dot{0}\;5\;{}_12\;6\;3\;{}_11\;{}_15\;{}_17\;{}_18$$
$$9\;{}_14\;7\;{}_13\;{}_16\;8\;4\;2\;\dot{1}$$

$$\frac{11}{19} = 0.{}_1\dot{5}\;{}_17\;{}_18\;9\;{}_14\;7\;{}_13\;{}_16\;8$$
$$4\;2\;1\;{}_10\;5\;{}_12\;6\;3\;{}_1\dot{1}$$

$$\frac{17}{29} = 0.{}_2\dot{5}\;{}_18\;6\;2\;{}_20\;{}_26\;{}_28\;{}_19\;{}_16\;{}_15\;5\;{}_21\;7\;{}_12$$
$$4\;{}_11\;{}_23\;{}_27\;9\;3\;1\;{}_10\;{}_13\;{}_14\;{}_24\;8\;{}_22\;{}_1\dot{7}$$

In each case the total of every column is the same: every column adds up to 9. This means that when we have got half way through a sum we can write down the second half from the first half: we just take every figure in the first half from 9 and this gives us the second half!

Now you may ask "how do I know when I am half way through?" and there is a simple way to find out. Look at the fraction. Take the numerator from the denominator: 19 – 1 = 18.
When 18 comes up you are half way through. In your recurring decimal for $\frac{1}{19}$ you should see 18 at the end of the first line!

Similarly for $\frac{11}{19}$, since 19–11 = 8 you should find 8 at the end of your first line.
And for $\frac{17}{29}$, 29–17 = 12 (written as ${}_12$) which comes up after 14 figures.

> If the difference of numerator and denominator comes up in a recurring decimal you are half way through and you can get the second half by taking all the figures in the first half from 9.
> (The difference of numerator and denominator does not always appear however).

If fact, if you include the remainder digits (the subscripts) each column adds up to the denominator.

15 Convert $\frac{9}{39}$ to a decimal.

First we note that the Ekadhika is **4** now (1 more than 3); we can write **E = 4**.
Also the half way number is 30 (39–9=30): **H = 39–9= 30**.
We begin by dividing the Ekadhika, **4**, into the numerator, 9: $\frac{9}{39} = 0.{}_12 3{}_30$
The first three steps give us 12, 3 and 30 and so we find that the half way number has come up after just 3 figures.
So we just take each of the first three figures from 9 to get the full answer:

$$\frac{9}{39} = 0.{}_1\dot{2}3{}_3076\dot{9}$$

\# You can of course find $\frac{9}{39}$ without using the half way number: just keep dividing by **4**.

❋ Find the recurring decimal for $\frac{10}{39}$. (E=4 and H=29)
You should get a 6-figure recurring decimal for this just as you did in Example 14, but in this case the half way number does not come up.

Answer: $\frac{10}{39} = 0.{}_2\dot{2}{}_2\,5{}_16\,4\,1{}_1\dot{0}$

❋ Plot the patterns for $\frac{9}{39}$ and $\frac{10}{39}$ on 9-point circles.

✎ **Practice D** Find the recurring decimal for:

a $\frac{25}{29}$ **b** $\frac{24}{39}$ **c** $\frac{29}{39}$ **d** $\frac{3}{49}$ **e** $\frac{44}{69}$ **f** $\frac{44}{79}$

g $\frac{1}{99}$ **h** $\frac{1}{9}$ (E will be 1 here because there is zero before the 9)

i plot your answers from **b** to **f** above on 9-point circles.

---

a $\frac{25}{29} = 0.{}_1\dot{8}62{}_20{}_26{}_28{}_19{}_16{}_155{}_217{}_124/137931034482 7\dot{5}$ **b** $\frac{24}{39} = 0.\dot{6}\,{}_21{}_15/38\dot{4}$ **c** $\frac{29}{39} = 0.{}_1\dot{7}\,{}_14{}_23{}_35{}_38{}_2\dot{9}$

d $\frac{3}{49} = 0.{}_3\dot{0}6{}_11{}_12{}_22{}_24{}_44{}_48{}_39{}_47{}_29{}_459{}_41{}_18{}_33{}_36{}_17{}_23{}_34{}_46/938775510204081632 65\dot{3}$

e $\frac{44}{69} = 0.{}_2\dot{6}\,{}_53{}_47{}_568{}_11{}_41{}_65{}_29{}_142{}_20{}_62{}_68{}_59{}_38{}_355{}_50{}_17{}_32{}_4\dot{4}$

f $\frac{44}{79} = 0.{}_4\dot{5}\,{}_55{}_76{}_49{}_162{}_20{}_42{}_25{}_13{}_51{}_36{}_4\dot{4}$ **g** $\frac{1}{99} = 0.\dot{0}\dot{1}$ **h** $\frac{1}{9} = 0.\dot{1}$

"On seeing this kind of work actually being performed by the little children, the doctors, professors and other "big-guns" of mathematics are wonder struck and exclaim: "Is this mathematics or magic?" And we invariably answer and say: "It is both. It is magic until you understand it; and it is mathematics thereafter"; and then we proceed to substantiate and prove the correctness of this reply of ours!"
From "Vedic Mathematics", Page xvii.

4.3c PROPORTIONATELY

So far all our fractions have had 9 as the last figure of the denominator.
In fact the special method we have been using can only be applied when the denominator ends in 9. However there are various other devices we can apply so that the method will work with other denominators.
One of these is to use the *Proportionately* formula.

16 Find the recurring decimal for $\frac{7}{13}$.

The denominator does not end in 9 here, it ends in 3.
However we know we can multiply the top and bottom of a fraction by any number we like without changing its value.

To get a 9 in the denominator we can multiply the numerator and denominator of $\frac{7}{13}$ by 3 to get: $\frac{7}{13} = \frac{21}{39}$.

And we now find the decimal for $\frac{21}{39}$ exactly as before: $\frac{21}{39} = 0.{}_1\dot{5}{}_3\,3{}_1 8\,4\,6\,\dot{1}$

The half way number is 18, and it comes up after 3 figures.

❈ What will $\frac{1}{7}$ need to be multiplied by on top and bottom so that the denominator ends in 9?
Answer: 7

In the next exercise each fraction will need to be multiplied so that the denominator ends in a 9.

Practice E Convert to recurring decimals:

a $\frac{1}{7}$ **b** $\frac{2}{13}$ **c** $\frac{5}{23}$ **d** $\frac{17}{33}$ **e** $\frac{9}{11}$ **f** $\frac{3}{17}$

---

a $\frac{1}{7} = \frac{7}{49} = 0.{}_2\dot{1}{}_1 4{}_4 2\,8\,5\,\dot{7}$ b $\frac{2}{13} = \frac{6}{39} = 0.{}_2\dot{1}{}_1\,5{}_3 3/84\dot{6}$ c $\frac{5}{23} = \frac{15}{69} = 0.{}_1\dot{2}{}_5\,1{}_2 7{}_6 39{}_2 13{}_3 0{}_2 4{}_3 3{}_5 47826086956\dot{5}$

d $\frac{17}{33} = \frac{51}{99} = 0.\dot{5}\dot{1}$ e $\frac{9}{11} = \frac{81}{99} = 0.\dot{8}\dot{1}$ f $\frac{3}{17} = \frac{21}{119} = 0.{}_9\dot{1}\,{}_7 7{}_5 6{}_8 47{}_7 0{}_{10} 5{}_9 8/8235294\dot{1}$

Practice F Find correct to 4 decimal places:

a $\frac{18}{59}$ **b** $\frac{67}{89}$ **c** $\frac{100}{109}$ **d** $1\frac{3}{7}$ **e** $\frac{20}{13}$ **f** $\frac{99}{49}$

---

a **0.3051** b **0.7528** c **0.9174** d **1.4286** e **1.5385** f **2.0204**

4.3d LONGER NUMERATORS

We are not restricted to denominators where the numerator is a whole number between 0 and the denominator. This method can work for any numerator.

17 Find **1.23 ÷ 19**.

We find $\frac{1}{19}$ and simply add 2 when we deal with the tenths and add 3 when we deal with the hundredths:

$$\frac{1.23}{19} = 0.{}_{1}0\ 6{}_{1}4\ 7{}_{1}3\ldots$$

We begin with 2 divided into 1 (as with $\frac{1}{19}$) and put ${}_{1}0$.

Next instead of dividing 2 into 10 we divide 2 into 12 because we add on the 2 in 1.23. This gives 6 which we put down.

Then the 3 in 1.23 is added to the 6, so we divide 2 into 9 and put down ${}_{1}4$.

From here on we just divide by 2 as before: 2 into 14 is 7 etc.

This considerably opens up the range of application of this method.

18 Find **2345 ÷ 49**.

$$\frac{2345}{49} = {}_{3}4{}_{3}7.{}_{2}8{}_{3}5\ldots$$

The steps are: 23 ÷ 5 = 4 rem 3, put ${}_{3}4$.
Now we add the 4 in 2345 to 34 to get 38, 38 ÷ 5 = 7 rem 3, put ${}_{3}7$.
Now add the 5 in 2345 to 37 to get 42, 42 ÷ 5 = 8 rem 2, put ${}_{2}8$.
From here on we divide as normal: 28 ÷ 5 = 5 rem 3 etc.

19 Find **72.7 ÷ 29**.

$$\frac{72.7}{29} = {}_{1}2.{}_{2}4{}_{1}(10){}_{2}6... = 2.506...$$

Here we get 10 in the third place, which means that the next step is 20 ÷ 3 (equals 6 remainder 2), and also that we have to carry 1 to the left at the end so that the 4 in the answer becomes 5.

Alternatively, at the second step

put $14 \div 3 = 5 \text{ rem } \bar{1}$:

$$\frac{72.7}{29} = {}_{1}2.{}_{\bar{1}}5{}_{2}0{}_{2}6... = 2.506...$$

Then ${}_{\bar{1}}5 = \bar{5}$ and we add the 7 in the tenths place of the numerator to this to get 2: 2 ÷ 3 = 0 rem 2, put ${}_{2}0$. And so on.

Or, thirdly to avoid the last 7 in 72.7,

start with $72.7 = 73.\bar{3}$:

$$\frac{73.\bar{3}}{29} = {}_{1}2.5{}_{2}0{}_{2}6...$$

Practice G Find to 4 S.F.

a 5.67 ÷ 19 **b** 67.8 ÷ 29 **c** 555 ÷ 39 **d** 0.0135 ÷ 79

e 321 ÷ 13 **f** 33 ÷ 9.5 **g** 19.19 ÷ 59 **h** 18.88 ÷ 19

---

a **0.2984** **b** **2.338** **c** **14.23** **d** **0.0001709**
e **24.69** **f** **3.474** **g** **0.3253** **h** **0.9937**

"And, in some very important and striking cases, sums requiring 30, 50, 100 or even more numerous and cumbrous "steps" of working (according to the current Western methods) can be answered in a single and simple step of work by the Vedic method! And little children (of only 10 or 12 years of age) merely look at the sums written on the blackboard (on the platform) and immediately shout out and dictate the answers from the body of the convocation hall (or other venue of demonstration). And this is because, as a matter of fact, each digit automatically yields its predecessor and its successor! and the children have merely to go on tossing off (or reeling off) the digits one after another (forwards or backwards) by mere mental arithmetic (without needing pen or pencil, paper or slate etc)!"

From "Vedic Mathematics", Page xvii.

LESSON 5
AUXILIARY FRACTIONS

SUMMARY

In this chapter we will extend the methods of finding recurring decimals in various directions. The Auxiliary Fractions, of which there are two types, are a great help in this.

5.1 AUXILIARY FRACTIONS: FIRST TYPE

The Auxiliary Fraction helps us to do the calculation: it helps to get us started and also gives us the Ekadhika, which we use continually.

The first type involves replacing the denominator of a fraction by its Ekadhika.

1 For $\frac{17}{19}$ the Auxiliary Fraction (AF) is $\frac{1.7}{2}$.

The AF is found by replacing the denominator by its Ekadhika (One More than the One Before) and dividing the numerator by 10.
So 19 is replaced by 2 and 17 is replaced by 1.7.

2 For $\frac{5}{28}$ the AF is $\frac{0.5}{3}$.

The 28 is replaced by *One More than the One Before*, which is 3, and the 5 is divided by 10 to give 0.5.

Practice A Write down the Auxiliary Fraction for each of the following:

a $\frac{33}{49}$ **b** $\frac{15}{79}$ **c** $\frac{53}{88}$ **d** $\frac{4}{37}$ **e** $\frac{8}{19}$ **f** $\frac{22}{139}$ **g** $\frac{1}{19}$

---

a $\frac{3.3}{5}$ b $\frac{1.5}{8}$ c $\frac{5.3}{9}$ d $\frac{0.4}{4}$ e $\frac{0.8}{2}$ f $\frac{2.2}{14}$ g $\frac{0.1}{2}$

Now we see how to use these to find recurring decimals.

3 Find the recurring decimal for $\frac{3}{7}$.

We know from previous work (see Section 4.3c) that we can do this by first obtaining a 9 in the denominator.
So we first multiply by 7 to get $\frac{3}{7} = \frac{21}{49}$. The AF for $\frac{21}{49}$ is $\frac{2.1}{5}$.

We then simply divide 5 into 2.1 in the usual way, and keep dividing by 5.
5 into 2.1 goes 0.4 with remainder 1 which is put before the 4:
AF = $\frac{2.1}{5}$ therefore $\frac{3}{7} = 0.{}_14$

Continuing we get $\frac{3}{7} = 0.{}_1\dot{4}{}_4 2{}_28\,5\,7\,\dot{1}$.

So the answer is $\frac{3}{7} = 0.\dot{4}2857\dot{1}$.

This is just the same as studied previously but with the additional help of the Auxiliary Fraction to get us started.

✎ **Practice B** Use Auxiliary Fractions to find recurring decimals for:

a $\frac{14}{39}$ **b** $\frac{6}{7}$ **c** $\frac{1}{13}$ **d** $\frac{21}{23}$

---

a $0.\dot{3}5897\dot{4}$ **b** $0.\dot{8}5714\dot{2}$ **c** $0.\dot{0}7692\dot{3}$ **d** $0.\dot{9}13043478260869565217\dot{3}$

5.2 DENOMINATORS ENDING IN 8, 7, 6

If a denominator ends not with a 9 but with a number close to it (8, 7 or 6) we can use the same method as for denominators ending in 9 but multiply the last figure at each step by 2, 3 or 4 before dividing.

4 $\frac{7}{48} = \mathbf{0.{}_21{}_24{}_35\,8{}_1\dot{3}}$.

The AF is $\frac{0.7}{5}$ (because 48 is still close to 50) so we begin by dividing 5 into 0.7.
This gives us 0.1 remainder 2: $\frac{7}{48} = 0.{}_21$

So far this is just what we would have done for $\frac{7}{49}$. But from now on, before we divide by 5 we must **double the last figure**, because 48 is **2** below 50.

Looking at ${}_21$ we double the last figure and therefore divide 5 into 22 rather than 21.
This gives 4 remainder 2, written ${}_24$ so we now have: $\frac{7}{48} = 0.{}_21{}_24$

We again double the last figure and divide 5 into 28 to get $_{3}5$.
Then doubling the 5 in 35 we divide 5 into 40, and so on.

Note how the 3 repeats itself at the end: we keep dividing 5 into 16 and get $_{1}3$ over and over again.

We can summarise the whole process as follows:

1. Find the Auxiliary Fraction. E.g. $\frac{0.7}{5}$ in the example above.

2. Divide the bottom of the AF into the top, writing any remainder **before** the answer figure.
E.g. $0.{}_{2}1$

3. From now on **double** the last figure of the number pair before dividing.
E.g. 5 into 22 (not 21)

When a fraction is converted to a decimal the resulting decimal may terminate or it may not. (e.g. 3÷4 = 0.75 which terminates). If the division does not terminate three things can happen:

1. the decimal has a single recurring figure (e.g. 2÷3 = 0.6666... = $0.\dot{6}$)

2. the decimal has a block of figures recurring (e.g. 2÷7 = $0.\dot{2}8571\dot{4}$)
3. the decimal has some non-recurring figures followed by a recurring figure or a block of recurring figures (e.g. 7÷30 = $0.2\dot{3}$ or 15÷28 = $0.53\dot{5}7142\dot{8}$)

Every factor of 2 or 5 in the denominator of a fraction contributes one non-recurring figure to the decimal (but 2 and 5 together, i.e. a factor of 10, contribute one figure). So $\frac{1}{22}$ and $\frac{1}{15}$ will have one non-recurring figure and a recurring part.

Practice C Find the full recurring decimal for:

a $\frac{5}{38}$ **b** $\frac{5}{78}$ **c** $\frac{7}{18}$

d Find the first 5 figures (after the decimal point) of the decimal for $\frac{13}{28}$.

---

a $\mathbf{0.1\dot{3}1578947368421052\dot{6}}$ **b** $\mathbf{0.0\dot{6}4102\dot{5}}$ **c** $\mathbf{0.3\dot{8}}$ **d** **0.46428**

5 $\frac{78}{87} = 0.{}_6\dot{8}\,{}_3 9\,{}_3 6\,{}_3 5\; 5\,{}_6 1\; 7\,{}_3 2\; 4\,{}_3 1\,{}_6 3\,{}_6 7\; 9\; 3\; 1\,{}_3 0\,{}_3 3\,{}_3 4\,{}_6 4\; 8\,{}_6 2\,{}_3 7\,{}_6 5\,{}_3 8\; 6\; 2\,{}_6 0\,{}_6 \dot{6}$

The AF is $\frac{7.8}{9}$ so we start with 9 into 7.8 goes 0.8 remainder 6 giving: $0.{}_6 8$

We now see 68 but because 87 is 3 below 90 we must treble the last figure at each step. So the 8 in 68 becomes 24, and added to 60 makes 84.

Dividing 84 by 9 we get 9 remainder 3 giving: $0.{}_6 8\,{}_3 9$

We now have 39. We treble the 9 to 27 and add the 30 which gives 57.
Then 57 divided by 9 goes 6 remainder 3 giving $0.{}_6 8\,{}_3 9\,{}_3 6$ and so on.

Practice D Find the full recurring decimal for:

a $\frac{9}{37}$ **b** $\frac{34}{77}$ **c** Find the first 6 figures of the decimal for $\frac{22}{57}$

---

a $0.\dot{2}4\dot{3}$ **b** $0.\dot{4}4155\dot{8}$ **c** 0.385964

6 Find $\frac{9}{28}$ giving 7 figures after the decimal point.

In finding $\frac{9}{28}$ you arrive at ${}_1\mathbf{5}$ in the 7th decimal place.

If you continue from here you get 3 into 20 (because the 5 is doubled) which goes 6 remainder 2.

This gives us ${}_1\mathbf{5}\,{}_2\mathbf{6}$.
Then 3 into 32 gives a 2-figure answer (10 rem 2).

This means that the 6 in ${}_2\mathbf{6}$ needs to be a 7 because 1 will be carried back from the next place.

So we do not write 3 into 20 goes 6 remainder 2 but write 7 remainder $\bar{1}$.

So we now have ${}_1\mathbf{5}\,{}_{\bar{1}}\mathbf{7}$.

This works nicely because we double 7 to get 14 and add this to $\overline{10}$ to get 4.
Then 3 into 4 goes 1 remainder 1 and the decimal is starting to repeat.

This gives $\mathbf{0.3\ 2\,{}_1\dot{1}4\,{}_2 2\ 8\,{}_1 5\,{}_{\bar{1}}\dot{7}}$ as the full answer.

Practice E Find the first 9 figures of the decimal for:

a $\frac{20}{47}$ **b** $\frac{19}{67}$

---

a **0.425531914** b **0.283582089**

5.3 AUXILIARY FRACTIONS: SECOND TYPE

5.3a DENOMINATORS ENDING IN 1

So far all the denominators we have encountered have been just **below** a multiple of ten, like 29, 38, 47 etc.

The second type of auxiliary fractions helps us to handle fractions which are just **over** a multiple of ten.

> To get the Auxiliary Fraction for fractions in which the denominator ends in a 1
> the numerator and denominator are both reduced by 1
> and the top and bottom are divided by 10.

7 For $\frac{7}{31}$ the AF is $\frac{0.6}{3}$.

Reducing top and bottom by 1 gives $\frac{6}{30}$ then 6 and 30 are divided by 10.

8 For $\frac{1}{21}$ the AF is $\frac{0.0}{2}$.

Practice F Write down the Auxiliary Fraction for:

a $\frac{9}{41}$ **b** $\frac{3}{91}$ **c** $\frac{44}{71}$ **d** $\frac{1}{61}$ **e** $\frac{2}{11}$ **f** $\frac{32}{301}$

---

a $\frac{0.8}{4}$ b $\frac{0.2}{9}$ c $\frac{4.3}{7}$ d $\frac{0.0}{6}$ e $\frac{0.1}{1}$ f $\frac{3.1}{30}$

The Auxiliary Fraction, as before, helps to get us started with the decimal.
In fact the method is the same as before but with one important difference: **before dividing at each step we take the last figure of the number being divided from 9.**

9 $\frac{19}{31} = \mathbf{0.6\,1_2 2\,9\,0\,3\,2_1 2\ldots}$

The AF is $\frac{1.8}{3}$ and so we begin by dividing 3 into 18. This gives: 0.6

From now on we take each answer figure (but not any of the small remainder figures) from 9 before dividing.

So we take 6 from 9 to get 3. Then 3 divided by 3 gives **1**:	0.61
Take this 1 from 9 to get 8, 8÷3=**2** rem **2**, so we now have:	0.61_22
Take the last 2 in 22 from 9 to get 27, 27÷3=**9**:	0.61_229
Take 9 from 9 to get 0, 0÷3=**0**:	0.61_2290
Take 0 from 9 to get 9, 9÷3=**3**, and so on.	

❋ Check you agree with the figures shown above and continue the division. The decimal begins to recur after 15 figures.

Answer: $\mathbf{0.\dot{6}1290322580645\dot{1}}$

> To summarise:
> 1. Write down the Auxiliary Fraction.
> 2. Divide the bottom of the AF into the top and put down the result, prefixing any remainder.
> 3. Take the last figure of the last number pair from 9 and divide again.

✐ **Practice G** Find the full recurring decimal for:

a $\frac{13}{21}$ **b** $\frac{22}{31}$ **c** $\frac{1}{21}$ **d** $\frac{17}{41}$ **e** $\frac{8}{51}$ **f** $\frac{4}{91}$ **g** $\frac{1}{81}$

---

a $0.\dot{6}1904\dot{7}$ **b** $0.\dot{7}0967741935483\dot{8}$ **c** $0.\dot{0}4761\dot{9}$ **d** $0.\dot{4}146\dot{3}$

e $0.\dot{1}56862745098039\dot{2}$ **f** $0.\dot{0}4395\dot{6}$ **g** $0.\dot{0}1234567\dot{9}$

ALTERNATIVE METHOD

An alternative method that follows on nicely and logically from the method for dividing by 19 is available, but it leads to occasional negative digits in the answer.

The explanation for the method of finding $\frac{1}{19}$ given earlier leads to the following complementary method for $\frac{1}{21}$ (and in fact denominators ending in 1 generally), in which the denominator is 20+1 rather than 20–1.

In dividing 21 into multiples of 20 we get:

$$\begin{array}{r}1\text{r}\bar{1}\\ \hline 21)20\end{array} \qquad \begin{array}{r}2\text{r}\bar{2}\\ \hline 21)40\end{array} \qquad \begin{array}{r}3\text{r}\bar{3}\\ \hline 21)60\end{array} \qquad \text{etc.}$$

This means we can use the same method for $\frac{1}{21}$ as for $\frac{1}{19}$, but the remainders become negative:

10 So in the case of $\frac{5}{21}$ we divide 2 into 5 and put: $\frac{5}{21}=0._{1}2$

Now when we divide by 2 we divide, not into 12 but into $1\bar{2}$, which is 8 (i.e. the last figure put down is considered negative).

Then 2 into 8 gives: $\frac{5}{21}=0._{1}24$

Next we count this 4 as $\bar{4}$, and $\bar{4}\div 2=\bar{2}$: $\frac{5}{21}=0._{1}24\bar{2}$

Next change the sign of $\bar{2}$ and find 2÷2 =1: $\frac{5}{21}=0._{1}24\bar{2}1$

And so on: $\frac{5}{21}=0._{1}24\bar{2}1_{\bar{1}}0\,\bar{5}_{1}2\ldots=0.\dot{2}3809\dot{5}.$

Explanation of method of Example 9

This method is the same as that for the first type except that it involves reducing the numerator of a fraction by 1 and taking the last answer digit from 9 at each step after the first.

The ingenious technique here is to take one of the units in the numerator and express it as $0.\dot{9}$. This gives a series of nines so that every negative remainder is changed to a positive number when 9 is added to it.

In the case of $\frac{19}{31}$ this becomes $\frac{18.9999.....}{31}$ and when we divide 18 by 3 and put down 6, we know the remainder is $\bar{6}$ but this is added to the first 9 in 18.9999….. And so on.

5.3b DENOMINATORS ENDING IN 2, 3, 4

11 $$\frac{\mathbf{5}}{\mathbf{42}}=\mathbf{0.1\,_{3}\dot{1}\,_{1}9_{1}0_{3}4_{3}7_{1}\,\dot{6}\,_{3}1}$$

We simply follow the procedure for denominators ending in 1, but, as in the previous section we double (for denominators ending in 3 we treble, and so on) the last digit **before** taking the complement from 9.

Here AF = $\frac{0.4}{4}$ (5 is reduced by 1, the 2 is dropped, and numerator and denominator are divided by 10) so we start with 4÷4 = 1: $\frac{5}{42}=0.1$

Then double the 1 and take its complement, which gives 7 and 7÷4 = ${}_{3}1$: $\frac{5}{42} = 0.1{}_{3}1$

Doubling 1 and taking the complement gives 7 again, so 37÷4 = ${}_{1}9$: $\frac{5}{42} = 0.1{}_{3}1{}_{1}9$

Next we double 9 to get 18, take this from 9 to get $\bar{9}$. So we now have ${}_{1}\bar{9}$ which is 1,

and this we divide by 4: $\frac{5}{42} = 0.1{}_{3}1{}_{1}9{}_{1}0$

And so on.

12 $\frac{7}{32} = 0.2\,{}_{2}1\;9\;\bar{3}\;5{}_{\bar{1}}0{}_{\bar{1}}0\ldots = \mathbf{0.21875}$.

Here we have a terminating decimal (ending is a series of zeros).
Note the $\bar{3}$ that comes up in the 4th place. Following this, we double $\bar{3}$ to $\bar{6}$, take $\bar{6}$ from 9 to get 15, which we divide by 3.

13 $\frac{12}{53} = \mathbf{0.{}_{1}2\,{}_{3}2\,{}_{3}6\,{}_{1}4\,{}_{2}1\,{}_{1}5\,{}_{4}0\,{}_{4}9\,{}_{2}4\,{}_{2}3\ldots}$

This is very similar except that as 53 is 3 over 50 we **treble** the last answer digit before taking the complement.

14 $\frac{1234}{51} = \mathbf{{}_{1}2\;4.{}_{4}1\,{}_{3}9\ldots}$

Here note that there will be two figures before the decimal point and we start the calculation from $\frac{12}{51}$.

So 11÷5 = ${}_{1}2$: $\frac{1234}{51} = {}_{1}2$

Then we take the complement of 2 to get 17, but add to this the 3 in the numerator.

So 20÷5 = 4: $\frac{1234}{51} = {}_{1}2\;4$

Complement of 4 is 5, add the 4 in the numerator to get 9, 9÷5 = ${}_{4}1$.
And so on.

✎ **Practice H** Find the recurring decimal for:

a $\frac{19}{52}$ **b** $\frac{5}{22}$ **c** $\frac{89}{92}$ (5 figures) **d** $\frac{16}{43}$ (9 figures)

e $\frac{1}{63}$ **f** $\frac{232}{41}$ (5 figures) **g** $\frac{13.54}{52}$ (5 figures)

---

a $\mathbf{0.{}_{3}3{}_{3}6{}_{2}\dot{5}{}_{4}3{}_{3}8{}_{3}4{}_{1}6{}_{2}\dot{1}{}_{2}5}$ **b** $\mathbf{0.2{}_{1}\dot{2}{}_{1}\dot{7}}$ **c** $\mathbf{0.{}_{7}9{}_{7}6{}_{4}7{}_{8}3{}_{2}9}$ **d** $\mathbf{0.{}_{3}3{}_{2}7\,2{}_{3}0{}_{3}9\,3\,0{}_{1}2{}_{1}3}$

e $\mathbf{0.{}_{0}\dot{0}{}_{3}1\,6{}_{\bar{3}}\bar{1}\,\bar{3}\,\dot{3} = 0.\dot{0}1\,5\,8\,7\,\dot{3}}$ **f** $\mathbf{{}_{2}5.{}_{2}6{}_{3}5{}_{2}8{}_{1}5}$ **g** $\mathbf{0.{}_{2}2\,6{}_{1}0{}_{4}3{}_{3}8}$

5.4 WORKING 2, 3 ETC. FIGURES AT A TIME

It is possible to speed up the calculation even further by working with groups of 2, 3, 4, 5, 6 or more figures at a time.

15 $\frac{1}{199} = 0.{}_{1}0\,0/5\,0/2\,5/{}_{1}1\,2/5\,6/2\,8/1\,4/0\,7/{}_{1}0\,3\,....$

The AF is now $\frac{0.01}{2}$ and because we have two 9's in the denominator we can get the answer **two figures at a time**.

We start from the AF, with 2 into 1 goes **00** remainder **1**, so we put down:

$0.{}_{1}00$

This gives 100, and 2 into 100 goes **50**, which we put down: $0.{}_{1}00/50$

(Note each pair of answer figures are separated by an oblique line only to help show the method)

Then 2 into 50 is **25**: $0.{}_{1}00/50/25$

Then 2 into 25 is **12** remainder **1**, so we put ${}_{1}12$: $0.{}_{1}00/50/25/{}_{1}12$

Then 2 into 112 and so on.

We get the answer at least as fast as we can write it down.

16 $\frac{59}{133} = \frac{177}{399} = 0.{}_{1}4\,4/3\,6/0\,9/{}_{1}0\,2/{}_{2}2\,5/{}_{1}5\,6\,....$

Here we see 133 in the denominator and so we multiply the top and bottom of $\frac{59}{133}$ by 3 to get $\frac{177}{199}$ to get 9's in the denominator.

So the AF is $\frac{1.77}{4}$ and we therefore divide by 4, two figures at a time.

4 into 177 goes 44 remainder 1 etc.

17 $\frac{108}{2001} = 0.{}_{1}0\,5\,3/9\,7\,3/0\,1\,3/4\,9\,3\,....$

AF = $\frac{0.107}{2}$ (Note that we are using the second type of AF here so we reduce the numerator of the given fraction by 1) and we work with groups of 3 figures:

$107 \div 2 =$ **53** remainder **1** so we put down: $0.{}_1053$

Next take each of the figures 053 from 9 to get 946,
so 2 into 1946 = **973**: $0.{}_1053/973$

Then 973 becomes 026, and dividing by 2 we get **013**.
Etc.

Practice I Find the first 5 groups of digits in the following:

a $\frac{88}{199}$ **b** $\frac{45}{299}$ **c** $\frac{535}{4999}$ (groups of 3 here)

d $\frac{70}{233}$ (multiply by 3 first) **e** $\frac{57}{201}$ (groups of 2 and take from 9 after the first group)

f $\frac{19}{43}$ (multiply by 7 first) **g** $\frac{222}{19999}$

h $\frac{37}{198}$ (divide by 2 in groups of 2 and double the last 2 figures after the first step)

---

a **0.44/22/11/05/52** b **0.15/05/01/67/22** c **0.107/021/404/280/856**
d **0.30/04/29/18/45** e **0.28/35/82/08/95**
f **0.44/18/60/46/51** g **0.0111/0055/5027/7513/8756**
h **0.18/68/68/68/68**

Special numbers like 201, 399, 1001 which are not prime and are near a base can be very useful here.
For example $\frac{1}{67}$ can be written as $\frac{3}{201}$ giving the answer two figures at a time, or alternatively as $\frac{597}{39999}$ giving the answer 4 figures at a time. We will look further at these numbers in Lesson 8.

"There are methods whereby ... we can easily transform any miscellaneous or non-descript denominator in question – by simple multiplication etc., to the requisite standard form which will bring them within the jurisdiction of the Auxiliary Fractions hereinabove explained.

In fact, the very discovery of these Auxiliaries and of their wonderful utility in the transmogrification of frightful-looking denominators of vulgar fractions into such simple and easy denominator-divisors must suffice and prepare the scientifically-minded seeker after Knowledge, for the marvellous devices still further on in the offing"
From "Vedic Mathematics", Page 272.

LESSON 6
VERTICALLY AND CROSSWISE

SUMMARY

6.1 Fractions – Adding and Subtracting Fractions (also Multiplication and Division).

6.2 General Multiplication – Multiplying 2, 3, 4 figure numbers in one line from left to right or from right to left, including use of bar numbers and algebraic products.

6.1 FRACTIONS

6.1a ADDING & SUBTRACTING FRACTIONS

These are usually found to be very difficult as the method is complicated and hard to remember. But the *Vertically and Crosswise formula* gives the answer immediately.

1 Find $\frac{2}{3}+\frac{1}{7}$

We multiply crosswise and add to get the numerator: $2\times7 + 1\times3 = \mathbf{17}$,
then multiply the denominators to get the denominator: $3\times7 = \mathbf{21}$.
So $\frac{2}{3}+\frac{1}{7}=\frac{\mathbf{17}}{\mathbf{21}}$.

PROOF

For both fractions to have the common denominator of 21 we must multiply the top and bottom of $\frac{2}{3}$ by 7 and the top and bottom of $\frac{1}{7}$ by 3. So each numerator gets multiplied by the denominator of the other fraction and these are then added.

Algebraically: $\frac{a}{b}+\frac{c}{d} = \frac{ad+cb}{bd}$.

Exactly the same pattern can be used for algebraic fractions as is used for numerical fractions. It is often the case in conventional mathematics that the arithmetic and algebraic methods are quite different. But in the Vedic system they are the same: the methods of multiplication, division, square roots for example, of algebraic expressions, are the same as those used for arithmetic calculations.

2 $\mathbf{\frac{x}{6}+\frac{2x}{5}=\frac{5x+12x}{30}=\frac{17x}{30}}$.

3 Find **a** $\mathbf{\frac{6}{7}-\frac{1}{4}}$ **b** $\mathbf{4\frac{1}{3}-1\frac{2}{5}}$.

a This is the same except we cross-multiply and subtract rather than add:

$$\frac{6}{7}-\frac{1}{4}=\frac{6\times4-1\times7}{7\times4}=\mathbf{\frac{17}{28}}.$$

b $4\frac{1}{3}-1\frac{2}{5}=3\frac{1\times5-2\times3}{3\times5}=3\frac{\bar{1}}{15}=\mathbf{2\frac{14}{15}}$. Here we get a negative numerator, but it is easily dealt with by taking $\frac{1}{15}$ from one of the whole ones.

Alternatively, to avoid the minus number here, put both fractions into top-heavy form and subtract (or take one unit from the 4 and find $\frac{4}{3}-\frac{2}{5}$).

Practice A Combine the following, cancelling down or leaving as mixed numbers where necessary:

a $\frac{2}{5}+\frac{1}{4}$ **b** $\frac{3}{8}+\frac{2}{5}$ **c** $\frac{1}{2}+\frac{2}{5}$ **d** $1\frac{1}{3}+2\frac{1}{4}$ **e** $3\frac{3}{4}+2\frac{1}{3}$

f $\frac{3}{5}-\frac{2}{7}$ **g** $\frac{8}{9}-\frac{1}{2}$ **h** $\frac{3}{4}-\frac{1}{20}$ **i** $5\frac{3}{5}-2\frac{1}{2}$ **j** $10\frac{2}{3}-1\frac{4}{5}$

k $\frac{5}{12}+\frac{7}{18}$ **l** $\frac{x}{2}+\frac{x}{7}$ **m** $\frac{x}{y}+\frac{y}{x}$ **n** $\frac{4}{x}-\frac{5}{y}$

a $\frac{13}{20}$	**b** $\frac{31}{40}$	**c** $\frac{9}{10}$	**d** $3\frac{7}{12}$	**e** $6\frac{1}{12}$
f $\frac{11}{35}$	**g** $\frac{7}{18}$	**h** $\frac{7}{10}$	**i** $3\frac{1}{10}$	**j** $8\frac{13}{15}$
k $\frac{29}{36}$	**l** $\frac{9x}{14}$	**m** $\frac{x^2+y^2}{xy}$	**n** $\frac{4y-5x}{xy}$	

We may note here that fractions are often written horizontally, for example 2/3. This is more consistent with the ratio notation (2:3) and place value. If fractions are written in this way then crosswise and horizontally (see Example 1) becomes crosswise and vertically. So for $\frac{2}{3}+\frac{1}{7}$:

$$\begin{array}{c} 2\,/\,3 \\ \underline{1\,/\,7} \quad + \\ \underline{17\,/\,21} \end{array}$$

6.1b A SIMPLIFICATION

In the question **k** above the numbers were rather large and some cancelling had to be done at the end. Where the denominators of two fractions are not relatively prime the working can be simplified as shown in the next example.

4 The denominators in $\frac{5}{12}+\frac{7}{18}$ are not relatively prime: there is a common factor of 6. We divide both denominators by this common factor and put these numbers below the denominators: $\underset{(2)}{\frac{5}{12}} + \underset{(3)}{\frac{7}{18}} = \frac{5\times3+7\times2}{12\times3} = \mathbf{\frac{29}{36}}$.

So we put 2 and 3 below 12 and 18.
Then when cross-multiplying we use the 2 and 3 rather than the 12 and 18.
For the denominator of the answer we cross-multiply in the denominators:
either 12×3 or 18×2, both give 36.

Subtraction of fractions with denominators which are not relatively prime is done in just the same way, except we subtract in the numerator as before.

Practice B Use this simplification to add or subtract the following:

a $\frac{1}{3}+\frac{4}{9}$ **b** $\frac{3}{8}+\frac{1}{6}$ **c** $\frac{3}{5}+\frac{3}{10}$ **d** $\frac{5}{6}-\frac{3}{4}$

e $\frac{5}{6}+\frac{3}{4}$ **f** $\frac{5}{18}-\frac{1}{27}$ **g** $3\frac{3}{4}-1\frac{1}{8}$ **h** $\frac{7}{36}-\frac{11}{60}$

---

a $\frac{7}{9}$ **b** $\frac{13}{24}$ **c** $\frac{9}{10}$ **d** $\frac{1}{12}$

e $1\frac{7}{12}$ **f** $\frac{13}{54}$ **g** $2\frac{5}{8}$ **h** $\frac{1}{90}$

6.1c COMPARING FRACTIONS

Some times we need to know whether one fraction is greater or smaller than another, or we may have to put fractions in order of size.

5 Put the fractions $\frac{4}{5}, \frac{2}{3}, \frac{5}{6}$ in ascending order.

Looking at the first two fractions we cross-multiply and subtract as if we wanted to subtract the fractions. Since $4\times3 > 2\times5$ the first fraction must be greater.
Doing this with $\frac{2}{3}$ and $\frac{5}{6}$ we find that $2\times6 < 5\times3$, so $\frac{5}{6}$ is greater than $\frac{2}{3}$.

If we now cross-multiply $\frac{4}{5}$ with $\frac{5}{6}$ we find that $\frac{5}{6}$ is greater.

So in ascending order the fractions are: $\mathbf{\frac{2}{3}, \frac{4}{5}, \frac{5}{6}}$.

Practice C Put the following fractions in ascending order:

a $\frac{1}{3}, \frac{2}{5}$ **b** $\frac{3}{4}, \frac{8}{11}$ **c** $\frac{2}{3}, \frac{7}{12}, \frac{3}{4}$ **d** $\frac{5}{6}, \frac{5}{8}, \frac{6}{7}$

---

a $\frac{1}{3}, \frac{2}{5}$ **b** $\frac{8}{11}, \frac{3}{4}$ **c** $\frac{7}{12}, \frac{2}{3}, \frac{3}{4}$ **d** $\frac{5}{8}, \frac{5}{6}, \frac{6}{7}$

6.1d MULTIPLICATION AND DIVISION

6 Find $\mathbf{\frac{1}{2} \times \frac{3}{4}}$.

$\frac{1}{2} \times \frac{3}{4} = \frac{1 \times 3}{2 \times 4} = \frac{3}{8}$. We simply multiply the numerators to get the numerator of the answer, and multiply the denominators to get the denominator of the answer.

7 Find $\mathbf{\frac{3}{4} \div \frac{2}{5}}$.

$\frac{3}{4} \div \frac{2}{5} = \frac{3 \times 5}{2 \times 4} = \frac{15}{8} = 1\frac{7}{8}$. We simply cross-multiply and put the first product over the second product.

We can summarise the methods of adding, subtracting, multiplying and dividing fractions with the following simple patterns:

Addition **Subtraction** **Multiplication** **Division**

This shows a welcome unity in the method for the four operations.

"People who have practical knowledge of the application of the Sutras need not go in or the theory side of it at all. The actual work can be done. Tremendous time is saved. It is a saving not merely of time and energy and money, but more than all, I feel, it is saving the child from tears that very often accompany the study of mathematics.".
From "Vedic Metaphysics", Page 170.

6.2 GENERAL MULTIPLICATION

6.2a REVISION

We have seen various methods of multiplication but they were all for special cases, where some special condition was satisfied: both numbers being close to 100 for example.

We come now to the general multiplication technique, under the *Vertically and Crosswise formula,* by which any two numbers can be multiplied together in one line, by mere mental arithmetic.

If you wish to tackle **written** left to right multiplications before (or instead of) Sections 6.2b, c, e, f (which show left to right multiplications done mentally) note the patterns shown in those sections and go to Section 6.2g.

If you wish to do right to left calculations first go straight to Section 6.2h (but you will need to refer back to the patterns shown in the previous sections) and Section 6.2d.

First let us briefly revise how we multiply mentally by a single figure number.

8 **74 × 8**.

We multiply each of the figures in 74 by 8 starting on the left:
7× 8 = **56** and 4 × 8 = **32**

These are combined by carrying the 3 in 32 over to the 6 in 56: $56,32 = 59$.

The inner figures are merged together. So **74 × 8 = 592**.

9 **827 × 3**.

The three products are **24**, **6**, **21**.
The first two products are combined: 24,6 = 246 no carry here as 6 is a single figure.
Then 246 is combined with the 21: $246,21 = 2481$. So **827 × 3 = 2481**.

10 **77 × 4**.

The products are **28**, **28**.
And $28,28 = 308$ (the 28 is increased by 2 to 30). So **77 × 4 = 308**.

Practice D Multiply the following mentally:

a 73 × 3 **b** 63 × 7 **c** 424 × 4 **d** 777 × 3

e 654 × 3 **f** 717 × 8 **g** 876 × 7

---

a 219 **b** 441 **c** 1 696 **d** 2 331
e 1 962 **f** 5 736 **g** 6 132

6.2b MULTIPLYING TWO-FIGURE NUMBERS

11 **21 × 23.**

Think of the numbers set out one below the other:

$$\begin{array}{cc} 2 & 1 \\ \underline{2} & \underline{3}\ \times \\ \hline \end{array}$$

There are 3 steps:

A. **multiply vertically** in the left-hand column: 2 × 2 = **4**, so 4 is the first figure of the answer.

$$\begin{array}{cc} 2 & 1 \\ | & \\ \underline{2} & \underline{3}\ \times \\ \mathbf{4} & \end{array}$$

B. **multiply crosswise** and add: 2 × 3 = 6, 1 × 2 = 2, 6 + 2 = **8**, so 8 is the middle figure of the answer.

$$\begin{array}{cc} 2 & 1 \\ & \times \\ \underline{2} & \underline{3}\ \times \\ \mathbf{4} & \mathbf{8} \end{array}$$

C. **multiply vertically** in the right-hand column: 1 × 3 = **3**, 3 is the last figure of the answer.

$$\begin{array}{ccc} 2 & & 1 \\ & & | \\ \underline{2} & & \underline{3}\ \times \\ \mathbf{4} & \mathbf{8} & \mathbf{3} \end{array}$$

12

$$\begin{array}{ccc} \mathbf{1} & & \mathbf{4} \\ \underline{\mathbf{2}} & & \underline{\mathbf{1}}\ \times \\ \mathbf{2} & \mathbf{9} & \mathbf{4} \end{array}$$

A. vertically on the left: 1 × 2 = **2**,
B. crosswise: 1 × 1 = 1, 4 × 2 = 8 and 1 + 8 = **9**,
C. vertically on the right: 4 × 1 = **4**.

This is of course very easy and straightforward, and we should now practice this vertical and crosswise pattern to establish the method.

Practice E Multiply mentally:

a 2 2	**b** 2 1	**c** 2 1	**d** 2 2	**e** 6 1	**f** 3 2	**g** 3 1	**h** 1 3
3 1 ×	3 1 ×	2 2 ×	1 3 ×	3 1 ×	2 1 ×	3 1 ×	1 3 ×
___	___	___	___	___	___	___	___

a	**682**	**b**	**651**	**c**	**462**	**d**	**286**
e	**1 891**	**f**	**672**	**g**	**961**	**h**	**169**

The previous examples involved no carry figures, so let us consider this next.

13

2 3
4 1 ×
9 4 3

The 3 steps give us: $2 \times 4 = \mathbf{8}$,
$2 \times 1 + 3 \times 4 = \mathbf{14}$,
$3 \times 1 = \mathbf{3}$.

The 14 here involves a carry figure, so in building up the answer mentally from the left we merge these numbers as before.

The mental steps are: 8
8,14 = 94 (the 1 is carried over to the left)
94,3 = 943

So **23 × 41 = 943**.

14

2 3
3 4 ×
7 8 2

The steps are: 6
6,17 = 77
77,12 = **782**.

15

3 3
4 4 ×
1 4 5 2

The steps are: 12
12,24 = 144
144,12 = **1452**.

We can now multiply any two 2-figure numbers together in one line.

Practice F Multiply the following mentally:

a 21	**b** 23	**c** 24	**d** 22	**e** 22	**f** 31
47	43	29	28	53	36
g 22	**h** 31	**i** 44	**j** 33	**k** 33	**l** 34
56	72	53	84	69	42

m	33	**n**	22	**o**	34	**p**	51	**q**	35	**r**	55
	34		52		66		54		67		59
s	54	**t**	55	**u**	44	**v**	45	**w**	48	**x**	34
	64		63		81		81		72		19

a	**987**	b	**989**	c	**696**	d	**616**	e	**1 166**	f	**1 116**
g	**1 232**	h	**2 232**	i	**2 332**	j	**2 772**	k	**2 277**	l	**1 428**
m	**1 122**	n	**1 144**	o	**2 244**	p	**2 754**	q	**2 345**	r	**3 245**
s	**3 456**	t	**3 465**	u	**3 564**	v	**3 645**	w	**3 456**	x	**646**

You may have found in this exercise that you prefer to start with the crosswise multiplications, and put the left and right vertical multiplications on afterwards.

For **writing** these sums see Section 6.2g.
For right to left calculations see Section 6.2h.

EXPLANATION

It is easy to understand how this method works.
The vertical product on the right multiplies units by units and so gives the number of units in the answer. The crosswise operation multiplies tens by units and units by tens and so gives the number of tens in the answer. And the vertical product on the left multiplies tens by tens and gives the number of hundreds in the answer.

EXPLANATION OF EARLIER SPECIAL METHOD

We can now explain the special method of multiplication under *By One More than the One Before* from Section 4.1 for multiplying numbers like 72 × 78 in which the first figures are the same and the last figures add up to 10.

Using the present sutra for 72 × 78:

$$\begin{array}{r} 7\ 2 \\ \underline{7\ 8} \\ 5\ 6_{7}1_{1}6 \end{array}$$

We see that the cross-product is eight 7's and two 7's, that is ten 7's, or 70. The zero here ensures that the 2-digit product 2 × 8 = 16 can go straight into the last two places, and this will always happen when the conditions for this type of product are met. The 7 in 70 means an extra 7 in the left-hand product.

As the method of squaring numbers that end in 5 is a special case of the above, this can also be explained this way.

THE DIGIT SUM CHECK

This is also available for checking all these calculations.
For example to check **33 × 44 = 1452** we convert these three numbers to digit sums and get: 6 × 8 = 3. This is correct in digit sum arithmetic as 6 × 8 = 48 and the digit sum of 48 is 3.

MULTIPLYING 3-FIGURE NUMBERS

16 **123 × 132 = 16236**.

The *Vertically and Crosswise* formula can be extended to deal with this, but in fact the previous vertical/crosswise/vertical pattern can be used on this sum also.

We can split the numbers up into 12/3 and 13/2, treating the 12 and 13 as if they were single figure numbers:

```
 12   3      vertically   12 × 13 = 156,
 13   2      crosswise    12 × 2 + 3 × 13 = 63,
162  3 6     vertically   3 × 2 = 6.
```

Combining these mentally we get:
156
156,63 = 1623,
1623,6 = **16236**

Practice G Multiply, treating the numbers as 2-figure numbers:

a	b	c	d	e
1 1 2	1 2 3	1 2 3	1 1 2	4 2 1
2 0 3	1 3 1	1 2 2	1 2 3	2 2

a 22 736 **b** 16 113 **c** 15 006 **d** 13 776 **e** 9 262

17 **304 × 412 = 125248**.

Here we may decide to split the numbers after the first figure: 3/04 × 4/12.

```
 3    04
 4    12
12  52  48
```

When we split the numbers in this way the answer appears **two digits at a time**.*

The three steps of the pattern are: 3×4 = **12**,
3×12 + 4×4 = **52**,
4×12 = **48**.

These give the three pairs of figures in the answer.

* This is because the left vertical product multiplies hundreds by hundreds and so gives the number of ten thousands in the answer. Similarly the crosswise step multiplies units by hundreds and so gives the hundreds figure (not the tens).

Practice H Multiply using pairs of digits:

a 211 × 304	**b** 307 × 407	**c** 203 × 432	**d** 211 × 311	**e** 504 × 504
f 501 × 501	**g** 712 × 112	**h** 703 × 211		

---

a **64 144**	b **124 949**	c **87 696**	d **65 621**	e **254 016**
f **251 001**	g **79 744**	h **148 333**		

6.2c MOVING MULTIPLIER

In multiplying a long number by a single figure, for example 4321 × 2, we multiply each of the figures in the long number by the single figure.
We may think of the 2 moving along the row, multiplying each figure vertically by 2 as it goes.

18 Find **4321 × 32**.

4 3 2 1 **3 2**	Similarly here we put 32 first of all at the extreme left. Then vertically on the left, 4 × 3 = **12**. And crosswise, 4×2 + 3×3 = **17**.
4 3 2 1 **3 2**	Then move the 32 along and multiply crosswise: 3×2 + 2×3 = **12**.
4 3 2 1 **3 2**	Moving the 32 once again: multiply crosswise, 2×2 + 1×3 = **7**. Finally the vertical product on the right is 1×2 = **2**.

These 5 results (in bold), 12,17,12,7,2 are combined mentally, as they are obtained, in the usual way:

12,17 = 137

137,12 = 1382

1382,7,2 = **138272**

So we multiply crosswise in every position, but we multiply vertically also at the very beginning and at the very end.

19 Find **31013 × 21**.

Here the 21 takes the positions:

3 1 0 1 3	**3 1 0 1 3**	**3 1 0 1 3**	**3 1 0 1 3**
2 1	** 2 1**	** 2 1**	** 2 1**

The 6 mental steps give: 6,5,1,2,7,3 so the answer is **651273**.

✎ **Practice I** Multiply using the moving multiplier method:

a 3 2 1	**b** 3 2 1	**c** 4 2 1	**d** 3 2 1	**e** 1 2 1 2
2 1	2 3	2 2	4 1	2 1
f 1 3 3 1	**g** 1 3 1 3	**h** 1 1 2 2 1	**i** 3 4 5 2 6	
2 2	3 1	2 2	1 1	

a	6 741	b	7 383	c	9 262	d	13 161	e	25 452
f	29 282	g	40 703	h	246 862	i	379786		

6.2d ALGEBRAIC MULTIPLICATIONS

The same Vertical and Crosswise pattern can be used to find the product of two binomials.

20 Multiply: **(2x + 5)(3x + 2)**.

2x	+ 5	Vertically on the left: 2x×3x=$\mathbf{6x^2}$.
3x	+ 2	Crosswise: 4x+15x=**19x**.
$\mathbf{6x^2 + 19x + 10}$		Vertically on the right: 5×2=**10**.

✎ **Practice J** Multiply:

a (x + 3)(x + 5) **b** (x + 7)(x – 2) **c** (x – 4)(x + 5)

d (x – 5)(x – 4) **e** (2x – 3)(3x + 6) **f** (3x – 1)(x + 7)

g (4x + 3)(2x – 5) **h** (x + 1)(9x – 1) **i** (2x + 1)(2x – 1)

a	$x^2+8x+15$	b	$x^2+5x-14$	c	x^2+x-20
d	$x^2-9x+20$	e	$6x^2+3x-18$	f	$3x^2+20x-7$
g	$8x^2-14x-15$	h	$9x^2+8x-1$	i	$4x^2-1$

So, unlike the current system, the same method is used for algebraic products as for arithmetic ones.

THE DIGIT SUM CHECK

The algebraic form of the digit sum check can be used.
If, for example, we wanted to check Example 20 above: $\mathbf{(2x + 5)(3x + 2)} = \mathbf{6x^2 + 19x + 10}$
we check that the product of the sum of the coefficients in the brackets on the left-hand side equals the sum of the coefficients on the right-hand side.

That is $(2 + 5)(3 + 2) = 6 + 19 + 10$.
Since both sides come to 35 this confirms the answer.

6.2e THREE-FIGURE NUMBERS

We can extend the Vertical and Crosswise multiplication to products of any size. The answer can always be found in one line.

21 Find **504 × 321**.

```
  5 0 4
  3 2 1
 ------
 161784
```

The extended pattern for multiplying 3-figure numbers is as follows.

A Vertically on the left, 5×3 = **15**.

```
 5 0 4
 |
 3 2 1
 -----
 15
```

B Then crosswise on the left,
5×2 + 0×3 = 10.
Combining the 15 and 10 as before:
15,10 = **160**.

```
 5 0 4
  ×
 3 2 1
 -----
 160
```

C Next we take 3 products and add them up,
5×1 + 0×2 + 4×3 = 17. And 160,17 = **1617**.

(actually we are gathering up the hundreds by multiplying hundreds by units, tens by tens and units by hundreds)

```
 5 0 4
  \|/
  /|\
 3 2 1
 -----
 1617
```

D Next we multiply crosswise on the right,
0×1 + 4×2 = 8: 1617,8 = **16178**.

```
 5 0 4
    ×
 3 2 1
 -----
 16178
```

E Finally, vertically on the right,
4×1 = 4: 16178,4 = **161784**.

```
 5 0 4
     |
```

3 2 1
161784

Note the symmetry in the 5 steps:
first there is 1 product, then 2, then 3, then 2, then 1.

We may summarise these steps as follows:

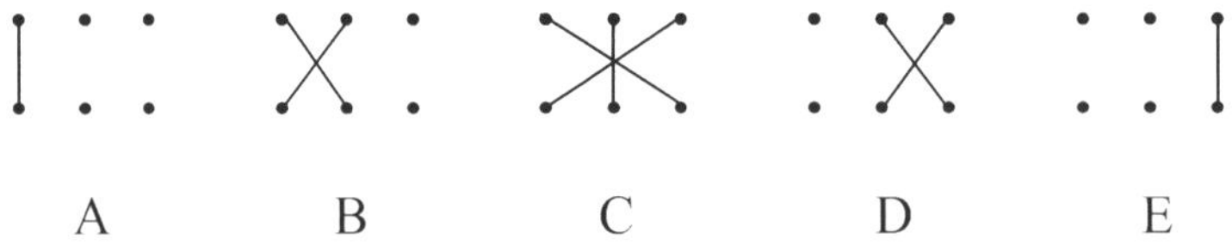

22

3 2 1
3 2 1 ×
103041

The five results are 9,12,10,4,1.
The mental steps are 9
9,12 = 102
102,10 = 1030
1030,4,1 = 103041.

23 Find **123 × 45**.

This can be done with the moving multiplier method or by the smaller vertical and crosswise pattern, treating 12 in 123 as a single digit .
Alternatively, we can put 045 for 45 and use the extended vertical and crosswise pattern:

1 2 3
0 4 5
5535

For the 5 steps we get 0,4,13,22,15.
Mentally we think 4; 53; 552; 5535.

Similarly we can multiply trinomials etc.

24 $\mathbf{(2x^2 + 3x + 4)(x^2 + 6) = 2x^4 + 17x^3 + 37x^2 + 46x + 24}$.

This is just like multiplying two 3-figure numbers.

$$\begin{array}{r} 2x^2 + 3x + 4 \\ x^2 + 0x + 6 \\ \hline 2x^4 + 3x^3 + 16x^2 + 18x + 24 \end{array}$$

Note that we put 0x for the missing x-term.

We multiply vertically on the left: $2x^2 \times x^2 = 2x^4$.
Crosswise in the first two columns: $(2x^2 \times 0x) + (3x \times x^2) = 3x^3$.
Next we have: $(2x^2 \times 6) + (3x \times 0x) + (4 \times x^2) = 12x^2 + 0x^2 + 4x^2 = 16x^2$.
Looking at the last two columns: $(3x \times 6) + (4 \times 0x) = 18x$.
And vertically on the right: $4 \times 6 = 24$.

The five steps automatically gather up the same powers of x.

✎ **Practice K** Multiply (there are no carries in the first few sums):

a 1 2 1 × 1 3 1	**b** 1 3 1 × 2 1 2	**c** 1 2 1 × 2 2 2	**d** 3 1 3 × 1 2 1	**e** 2 1 2 × 3 1 3	**f** 1 2 3 × 3 2 1
g 2 1 2 × 4 1 4	**h** 2 2 2 × 3 3 3	**i** 2 4 6 × 3 3 3	**j** 1 0 5 × 5 0 7	**k** 1 0 6 × 2 2 2	**l** 5 1 5 × 5 5 5
m 4 4 4 × 7 7 7	**n** 3 2 1 × 3 2 1	**o** 1 2 3 × 2 7 1	**p** 1 2 4 × 3 5 6	**q** 1 3 7 × 8 0 3	**r** 1 3 1 × 7 7 1

s $(2x^2 + 2x + 3)(3x^2 + 5x + 1)$ **t** $(x^2 + 6x - 3)(2x^2 + 3x + 4)$

u $(2x^2 + 2xy + 3y^2)(x^2 + 5xy + y^2)$ **v** $(5x^2 - 3x - 8)(5x + 2)$

---

a	**15 851**	b	**27 772**	c	**26 862**	d	**37 873**	e	**66 356**	f	**39 483**
g	**87 768**	h	**73 926**	i	**81 918**	j	**53 235**	k	**23 532**	l	**285 825**
m	**344 988**	n	**103 041**	o	**33 333**	p	**44 144**	q	**110 011**	r	**101 001**

s **$6x^4 + 16x^3 + 21x^2 + 17x + 3$** t **$2x^4 + 15x^3 + 16x^2 + 15x - 12$**

u **$2x^4 + 12x^3y + 15x^2y^2 + 17xy^3 + 3y^4$** v **$25x^3 - 5x^2 - 46x - 16$**

6.2f FOUR-FIGURE NUMBERS

Once the vertical and crosswise method is understood it can be extended to multiply numbers of any size. We here extend the pattern one stage further, and multiply two 4-figure numbers.

25

$$\begin{array}{r} 3\ 2\ 0\ 1 \\ \underline{4\ 3\ 0\ 2} \times \\ \underline{1\,3\,7\,7\,0\,7\,0\,2} \end{array}$$

"We thus follow a process of ascent and descent (going forward with the digits on the upper row and coming rearward with the digits on the lower row)."
From "Vedic Mathematics", Page 42.

The 7 steps are illustrated as follows:

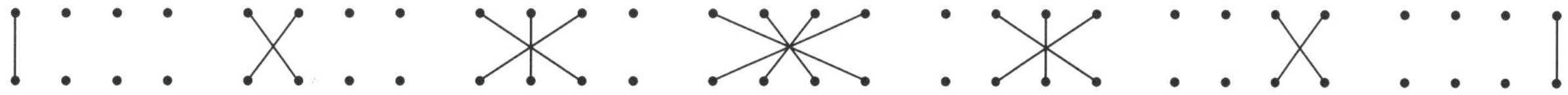

A B C D E F G

Working from left to right we get:
A. 3×4 = 12
B. 3×3 + 2×4 = 17
C. 3×0 + 2×3 + 0×4 = 6
D. 3×2 + 2×0 + 0×3 + 1×4 = 10
E. 2×2 + 0×0 + 1×3 = 7

F. $0\times2 + 1\times0 = 0$
G. $1\times2 = 2$

The mental steps are therefore: 12,17 = 137

137,6 = 1376

1376,10 = 13770

13770,7,0,2 = **13770702**

✐ **Practice L** Multiply the following from left to right or from right to left:

a 2 1 3 1	**b** 2 0 2 1	**c** 3 2 0 1	**d** 5 1 1 3
3 0 2 2	1 1 2 2	4 0 1 2	5 3 3 1

a 6,439,882 b 2,267,562 c 12,842,412 d 27,257,403

6.2g WRITING LEFT TO RIGHT SUMS

This is a continuation of the method introduced in Section 1.3 and will be found useful later for doing combined operations .

26 **63 × 74 = 662**.

$$\begin{array}{r} 6\ 3 \\ 7\ 4 \\ \hline 4_{2}6_{5}6\ 2 \end{array}$$

6 × 7 = 42, put down **4** and carry **2**;
(6 × 4) + (3 × 7) = 45, add carried 2, **as 20**,
45 + 20 = 65, put down **6**, carry **5**;
3 × 4 = 12, add carried 5, **as 50**,
12 + 50 = 62, put down **62**.

27 **61 × 83 = 5063**.

$$\begin{array}{r} 6\ 1 \\ 8\ 3 \\ \hline 4_{8}{}^{1}0_{6}6\ 3 \end{array} = \mathbf{5063}$$

Here if we start with 4_8 we get 106 next and have to carry 1 to the left.

$$\begin{array}{r} 6\ 1 \\ 8\ 3 \\ \hline 5_{\bar{2}}0_{6}6\ 3 \end{array}$$

Alternatively, if we put $5_{\bar{2}}$ instead of 4_8 we get 06 next.

28

$$\begin{array}{r} 4\ 3\ 2\ 1 \\ 3\ 2 \\ \hline \end{array}$$

The moving multiplier method can be used for sums like this.

$\underline{1_2 3_7 8_2 2_7 7\ 2}$

29 **243 × 316 = 76788**.

$$\begin{array}{r} 2\ 4\ 3 \\ \underline{3\ 1\ 6} \\ \underline{0_6 7_4 6_5 7_7 8\ 8} \end{array}$$

2 × 3 = 6, put down **0** and carry **6**, etc.

30 **3201 × 4302 = 13770702**.

$$\begin{array}{r} 3\ 2\ 0\ 1 \\ \underline{4\ 3\ 0\ 2} \\ \underline{1_2 3_7 7_6 7_0 0_7 7_0 0\ 2} \end{array}$$

Practice M

a 47 × 36 **b** 62 × 71 **c** 56 × 65 **d** 41 × 14 **e** 73 × 84

f 35 × 23 **g** 3434 × 42 **h** 567 × 82 **i** 444 × 321 **j** 765 × 321

k 357 × 223 **l** 135 × 531 **m** 5103 × 3221

---

a $1_2 6_5 92$ b $4_2 4_0 02$ c $3_0 6_1 40$ d $0_4 5_7 74$ e $6_{\bar{4}} 1_2 32$ or $5_6{}^1 1_2 32 = 6132$

f $0_6 8_{\bar{1}} 05$ or $0_6 7_9{}^1 05 = 805$ g 144228 h 46494 i $1_2 4_0 2_4 5_2 24$ j $2_1 4_2 5_4 5_6 65$

k $0_6 7_6 9_3 6_{\bar{1}} 11$ l $0_5 7_{\bar{2}} 1_5 6_8 85$ m 16436763

6.2h FROM RIGHT TO LEFT

We can also calculate from right to left if we prefer, though for mental calculations left to right is better.

31 **86 × 23 = 1978**.

$$\begin{array}{r} 8\ 6 \\ \underline{2\ 3} \\ \underline{1\ 9_3 7_1 8} \end{array}$$

The steps are:
1) 6×3 = 18,
2) (8×3) + (6×2) = 36, 36 + 1 = 37,
3) 8×2 = 16, 16 +3 = 19.

32 **4321 × 32 = 138272**.

$$\begin{array}{cccc} 4&3&2&1\\ &&3&2 \end{array} \qquad \begin{array}{cccc} 4&3&2&1\\ &3&2& \end{array} \qquad \begin{array}{cccc} 4&3&2&1\\ 3&2&& \end{array}$$

The 32 takes the three positions shown (moving multiplier method – see Section 6.2c).

$$\begin{array}{cccccc} &&4&3&2&1\\ &&&&3&2\\ \hline 1&3&{}_18&{}_12&7&2 \end{array}$$

The steps (starting at the right are:
1) 1×2 = 2,
2) 2×2 + 1×3 = 7,
3) 3×2 + 2×3 = 12, write $_12$,
4) 4×2 + 3×3 = 17, 17 + carried 1 = 18, put $_18$,
5) 4×3 = 12, 12 + carried 1 = 13..

33

$$\begin{array}{ccccc} &&\mathbf{2}&\mathbf{3}&\mathbf{4}\\ &&\mathbf{2}&\mathbf{3}&\mathbf{4} \;\times\\ \hline \mathbf{5}&{}_1\mathbf{4}&{}_2\mathbf{7}&{}_2\mathbf{5}&{}_1\mathbf{6} \end{array}$$

We simply do the same operations (see Section 6.2e) but start at the right side:
4×4 = **16**, put down 6 and carry 1 to the left.
3×4 + 4×3 = 24, 24 + carried 1 = **25**, put down 5 and carry 2.
And so on.

34 Calculating Example 25 from right to left we get the numbers from G to A shown:

$$\begin{array}{ccccccc} &&&\mathbf{3}&\mathbf{2}&\mathbf{0}&\mathbf{1}\\ &&&\mathbf{4}&\mathbf{3}&\mathbf{0}&\mathbf{2}\\ \hline \mathbf{1}&\mathbf{3}&{}_1\mathbf{7}&\mathbf{7}&{}_1\mathbf{0}&\mathbf{7}&\mathbf{0}\;\mathbf{2} \end{array}$$

Practice N Multiply the following from right to left:

a 33 × 41 **b** 52 × 62 **c** 37 × 43 **d** 73 × 14

e 444 × 333 **f** 543 × 345 **g** 707 × 333 **h** 623 × 32

i 3231 × 2342

---

a 1353 **b 3224** **c 1591** **d 1022**
e 147,852 **f 187,335** **g 235,431** **h 19936**

i 7567002

SETTING THE SUMS OUT

In Example 33 each of the five steps had a center of symmetry.
The five dots on the right show these five centers and as we move from left to right or right to left through the sum it is as if there is a dot moving through the sum.

2 3 4
• • • • •
2 3 4 ×
54756

In the calculation shown here the units figure of the result of each of the five steps is placed under the dot for that step

Other ways of setting the sums and answers out are possible and may be preferred.

6.2i USING BAR NUMBERS

35 Find **34 × 19**.

This is the last sum from Practice F, but suppose here we would like to use bar numbers to remove the large digit 9.

So we change **34 × 19** to **34 ×** $\mathbf{2\bar{1}}$:

$$\begin{array}{cc} 3 & 4 \\ 2 & \bar{1} \\ \hline \end{array}$$
$$65\bar{4} = \mathbf{646}$$

So the bar numbers can be quite useful in removing all digits 6, 7, 8, and 9 from a calculation. This gives us a choice about how to do a sum (and in fact to facilitate cancellations of positive and negative digits it is sometimes best to leave some digits over 5 in a sum, but this needs some practice to get a feel for what works best).

Practice O

a 19 × 24 **b** 59 × 23 **c** 28 × 31 **d** 19 × 49 **e** 38 × 38

f 292 × 398 **g** 309 × 279

---

a 456 b 1357 c 868 d 931 e 1444
f 116216 g 86211

LESSON 7
SQUARES AND SQUARE ROOTS

SUMMARY

7.1 General Squaring – of numbers of any size and algebraic expressions.
7.2 Square Roots of Perfect Squares – square roots of numbers up to 40,000.

7.1 GENERAL SQUARING

The *Vertically and Crosswise* formula simplifies nicely when the numbers being multiplied are the same, and gives us a really easy method for squaring numbers.

7.1a TWO-FIGURE NUMBERS

We will use the term **Duplex**, D, as follows:

for 1 figure **D is its square**, e.g. $D(4) = 4^2 = \mathbf{16}$;

for 2 figures **D is twice their product**, e.g. $D(43) = 2\times4\times3 = \mathbf{24}$.

✎ **Practice A** Find the Duplex of the following numbers:

a 5 **b** 23 **c** 55 **d** 2 **e** 14 **f** 77 **g** 26 **h** 90

---

a 25 b 12 c 50 d 4 e 8 f 98 g 24 h 0

To do written squaring from left to right before or instead of the mental method note the technique shown in Sections 7.1a, b, d and go to Section 7.1e.

> The square of any number is just the total of its Duplexes, combined in the way we have been using for mental multiplication.

1 $\mathbf{43^2 = 1849}$.

Working from left to right there are three duplexes in 43: D(4), D(43) and D(3).
D(4) = **16**, D(43) = **24**, D(3) = **9**.

Combining these three results in the usual way we get:

16
16,24 = 184
184,9 = **1849**

2 **64^2 = 4096**.

D(6) = **36**, D(64) = **48**, D(4) = **16**,
So mentally we get:

36
36,48 = 408
408,16 = **4096**

Practice B Square the following:

a 31 **b** 14 **c** 41 **d** 26 **e** 23 **f** 32 **g** 21

h 66 **i** 81 **j** 91 **k** 56 **l** 63 **m** 77

a	**961**	b	**196**	c	**1681**	d	**676**	e	**529**	f	**1024**	g	**441**
h	**4356**	i	**6561**	j	**8281**	k	**3136**	l	**3969**	m	**5929**		

Proof: $(10a + b)^2 = 100(a^2) + 10(2ab) + b^2$

7.1b NUMBER SPLITTING

You may recall that we could sometimes group two figures as one when we were multiplying two 2-figure numbers together. This also applies to squaring.

3 **123^2 = 15129**.

Here we may think of 123 as 12/3, as if it were a 2-figure number:
D(12) = 12^2 = **144**,
D(12/3) = 2 × 12 × 3 = **72**,
D(3) = 3^2 = **9**.
Combining these: 144,72 = 1512, and 1512,9 = **15129**.

Practice C Square the following, grouping the first pair of figures together:

a 121 **b** 104 **c** 203 **d** 113 **e** 116 **f** 108

a	**14 641**	b	**10 816**	c	**41 209**	d	**12 769**	e	**13 456**	f	**11 664**

4 $\mathbf{312^2 = 97344}$.

Here we can split the number into 3/12 but we must work with **pairs of digits**:
D(3) = **9**, D(3/12) = **72**, D(12) = **144**.
Combining: 9,72 = 972 we can put both figures of 72 after the 9,
972,144 = **97344**.

Practice D Square the following, grouping the last 2 figures together:

a 211	**b** 412	**c** 304	**d** 902	**e** 407	**f** 222	**g** 711

a **44 521**	b **169 744**	c **92 416**	d **813 604**	e **165 649**	f **49 284**	g **505 521**

7.1c ALGEBRAIC SQUARING

This is just like squaring 2–figure numbers.

5 Find $\mathbf{(2x + 3)^2}$.

There are three Duplexes: D(2x) = $\mathbf{4x^2}$, D(2x+3) = $2\times 2x\times 3$ = **12x**, D(3) = **9**.
So $\mathbf{(2x + 3)^2 = 4x^2 + 12x + 9}$.

6 Find $\mathbf{(x - 3y)^2}$.

Similarly: D(x) = $\mathbf{x^2}$, D(x–3y) = $2\times x\times -3y$ = **–6xy**, D(–3y) = $\mathbf{9y^2}$.
So $\mathbf{(x - 3y)^2 = x^2 - 6xy + 9y^2}$.

Practice E Square the following:

a $(3x + 4)$	**b** $(5y + 2)$	**c** $(2x - 1)$	**d** $(x + 7)$	**e** $(x - 5)$	**f** $(x + 2y)$
g $(3x + 5y)$	**h** $(2a + b)$	**i** $(2x - 3y)$	**j** $(x + y)$	**k** $(x - y)$	**l** $(x - 8y)$

a $9x^2+24x+16$	b $25y^2+20y+4$	c $4x^2-4x+1$	d $x^2+14x+49$	e $x^2-10x+25$	f $x^2+4xy+4y^2$
g $9x^2+30xy+25y^2$	h $4a^2+4ab+b^2$	i $4x^2-12xy+9y^2$	j $x^2+2xy+y^2$	k $x^2-2xy+y^2$	l $x^2-16xy+64y^2$

7.1d SQUARING LONGER NUMBERS

We can also find the duplex of 3-figure numbers or bigger.

For 3 figures D is **twice the product of the outer pair + the square of the middle digit**,

e.g. $D(137) = 2\times1\times7 + 3^2 = \mathbf{23}$;

for 4 figures D is **twice the product of the outer pair + twice the product of the inner pair**,

e.g. $D(1034) = 2\times1\times4 + 2\times0\times3 = \mathbf{8}$;

$D(10345) = 2\times1\times5 + 2\times0\times4 + 3^2 = \mathbf{19}$;
and so on.

Practice F Find the duplex of the following numbers:

a 234	**b** 282	**c** 717	**d** 304	**e** 270
f 1234	**g** 3032	**h** 7130	**i** 20121	**j** 32104

---

a **25**	b **72**	c **99**	d **24**	e **49**
f **20**	g **12**	h **6**	i **5**	j **25**

Just as for 2-figure numbers the square of any number is just the total of its duplexes.

7 $\mathbf{341^2 = 116281}$.

Here we have a 3-figure number:
D(3) = 9, D(34) = 24, D(341) = 22, D(41) = 8, D(1) = 1.

Mentally 9,24 = 114

114,22 = 1162

1162,8,1 = 116281.

8 $\mathbf{4332^2 = 18766224}$.

D(4) = 16,	D(43) = 24,	D(433) = 33,	D(4332) = 34,
D(332) = 21,	D(32) = 12,	D(2) = 4.	

Mentally: 16,24 = 184

184,33 = 1873

1873,34 = 18764

18764,21 = 187661

187661,12 = 1876622

1876622,4 = 18766224.

Practice G Square the following numbers:

a 212 **b** 131 **c** 204 **d** 513 **e** 263 **f** 264

g 313 **h** 217 **i** 3103 **j** 2132 **k** 1414 **l** 4144

m Find x given that $\mathbf{x23^2 = 388129}$

n Find b, c and d given that $\mathbf{b15^2} = \mathbf{17cccd}$

a	**44 944**	b	**17 161**	c	**41 616**	d	**263 169**	e	**69 169**	f	**69 696**
g	**97 969**	h	**47 089**	i	**9 628 609**	j	**4 545 424**	k	**1 999 396**	l	**17 172 736**
m	**x=6**	n	**b=4, c=2, d=5**								

7.1e WRITTEN CALCULATIONS – LEFT TO RIGHT

In a similar way to Section 6.2g we can write down the answers step by step when squaring left to right.

9 $\mathbf{43^2} = 1_6 8_4 49 = \mathbf{1849}$.

The steps are: D(4) = 16, put $\mathbf{1}_6$.
D(43) = 24, 24 + 60 = 84, put $\mathbf{8}_4$.
D(3) = 9, 9 + 40 = 49, put **49**.

10 $\mathbf{234^2} = 0_4 5_2 4_5 7_4 56 = \mathbf{54756}$.

D(2) = 4, put 0_4.
D(23) = 12, 12 + 40 = 52, put $\mathbf{5}_2$.
D(234) = 25, 25 + 20 = 45, put $\mathbf{4}_5$.
D(34) = 24, 24 + 50 = 74, put $\mathbf{7}_4$.
D(4) = 16, 16 + 40 = 56, put **56**.

But note the carry to the left in the next example.

11 $\mathbf{191^2} = 0_1 2_8 {}^1 6_3 4_8 81 = \mathbf{36481}$

At the third step we get D(191) = 83, 83 + 80 = 163, put ${}^1 6_3$.
The 1 in the 16 here is carried leftwards to give **36**481.

Alternatively, when you get 28 at the second step, write it as $3\bar{2}$.

✎ **Practice H** Square the numbers in Practice G in this way.

7.1f WRITTEN CALCULATIONS – RIGHT TO LEFT

This is similar to Section 6.2h.

12 $\mathbf{43^2} = 18_2 49 = \mathbf{1849}$.

The steps are: D(3) = 9, put 9:	$\mathbf{43^2 = 9}$
D(43) = 24, put $_2 4$:	$\mathbf{43^2 = {}_2 49}$
D(4) = 16, 16 + carried 2 = 18, put 18:	$\mathbf{43^2 = 18_2 49}$

13 $\mathbf{234^2} = 5_1 4_2 7_2 5_1 6 = \mathbf{54756}$.

D(4) = 16:	$\mathbf{234^2 = {}_1 6}$
D(34) = 24, 24 + 1 = 25:	$\mathbf{234^2 = {}_2 5_1 6}$
D(234) = 25, 25 + 2 = 27:	$\mathbf{234^2 = {}_2 7_2 5_1 6}$
D(23) = 12, 12 + 2 = 14:	$\mathbf{234^2 = {}_1 4_2 7_2 5_1 6}$
D(2) = 4, 4 + 1 = 5:	$\mathbf{234^2 = 5_1 4_2 7_2 5_1 6}$

14 $\mathbf{191^2} = 3_2 6_8 4_1 81 = \mathbf{36481}$.

D(1) = 1:	$\mathbf{191^2 = 1}$
D(91) = 18:	$\mathbf{191^2 = {}_1 81}$
D(191) = 83, 83 + 1 = 84:	$\mathbf{191^2 = {}_8 4_1 81}$
D(19) = 18, 18 + 8 = 26:	$\mathbf{191^2 = {}_2 6_8 4_1 81}$
D(1) = 1, 1 + 2 = 3:	$\mathbf{191^2 = 3_2 6_8 4_1 81}$

An alternative method for 191^2 would be to remove the large digit, 9: $191^2 = 2\bar{1}1^2$.
And $2\bar{1}1^2 = 4\bar{4}5\bar{2}1$ (no carries!) $= 36481$.

✎ **Practice I** Square the numbers in Practice G from right to left.

7.2 SQUARE ROOTS OF PERFECT SQUARES

Here we look at the square root of perfect squares.

It is worth noting the following facts:

> Square numbers can only end in 1, 4, 5, 6, 9, 0
> and they can only have digit sums of 1, 4, 7, 9.

It is therefore often easy to tell if a number is not a perfect square.
(This can make a good lesson topic for pupils.)

15 Find $\sqrt{\mathbf{6889}}$.

First note that there are two groups of figures, 68'89, so we expect a 2-figure answer (we mark off pairs of digits from the right).

Next we use *The First by the First and the Last by the Last.* Looking at the 68 at the beginning we can see that since 68 is greater than 64 (8^2) and less than 81 (9^2) the first figure must be 8.

Or looking at it another way 6889 is between 6400 and 8100

$$6400 = 80^2$$
$$\mathbf{6889 = 8?^2}$$
$$8100 = 90^2$$

so $\sqrt{6889}$ must be between 80 and 90. I.e. it must be eighty something.
Now we look at the last figure of 6889, which is 9.
Any number ending with 3 will end with 9 when it is squared so the number we are looking for could be 83.

But any number ending in 7 will also end in 9 when it is squared so the number could also be 87.

So is the answer 83 or 87?
There are two easy ways of deciding. One is to use the digit sums.
If $87^2 = 6889$ then converting to digit sums we get $6^2 = 4$, which is not correct.
But $83^2 = 6889$ becomes $2^2 = 4$, so the answer must be **83**.

The other method is to recall that since $85^2 = 7225$ and 6889 is **below** this $\sqrt{6889}$ must be **below 85**. So it must be **83**.

> To find the square root of a perfect square we find the first figure by looking at the first figures of the square, and we find two possible last figures by looking at the last figure. We then decide which is correct either by considering the digit sums or by considering the square of their mean.

16 Find $\sqrt{\mathbf{5776}}$.

The 57 at the beginning is between 49 and 64, so the first figure must be 7.

The 6 at the end tells us the square root ends in 4 or 6.
So the answer is 74 or 76.

$74^2 = 5776$ becomes $4 = 7$ which is not true in terms of digit sums, so 74 is not the answer.
$76^2 = 5776$ becomes $7 = 7$ so **76** is the answer.

Alternatively to choose between 74 and 76 we note that $75^2 = 5625$ and 5776 is greater than this so the square root must be greater than 75. So it must be **76**.

Practice J Find (mentally) the square root of:

a 2116	**b** 5329	**c** 1444	**d** 6724	**e** 3481	**f** 4489	**g** 8836
h 361	**i** 784	**j** 3721	**k** 2209	**l** 4225	**m** 9604	**n** 5929

---

a 46	**b 73**	**c 38**	**d 82**	**e 59**	**f 67**	**g 94**
h 19	**i 28**	**j 61**	**k 47**	**l 65**	**m 98**	**n 77**

As you will have seen, square numbers ending in 5 must have a square root ending in 5: there is only one possibility for the last figure.

17 Find $\sqrt{\mathbf{31329}}$.

If we mark off pairs of digits from the right here we get 3'13'29.
We have three groups, indicating that the answer is a 3-figure number.

However if you know all the square numbers up to 20^2 we can still get the answer by this method. We think of 31329 as 313'29.

Since 313 lies between 289 (17^2) and 324 (18^2) the first two figures must be 17.
And the last figure is 3 or 7, so 173 and 177 are the two possibilities.

The digit sum will then confirm 177 as the right one.
Alternatively you may argue that since 313 is closer to 324 than 289 it will be 177 rather than 173.

✎ **Practice K** Find the square root of :

a 26896	**b** 32761	**c** 16129	**d** 24964	**e** 36864	**f** 18496
g 21025	**h** 29929	**i** 14161	**j** 11236		

---

a **164**	b **181**	c **127**	d **158**	e **192**	f **136**
g **145**	h **173**	i **119**	j **106**		

LESSON 8
SPECIAL MULTIPLICATION METHODS

SUMMARY

8.1 Special Numbers – finding all the factors of special numbers in a sum.
8.2 Using the Average – finding products by squaring their average.
8.3 Multiplication by Nines – multiplying by 9, 99, 999 etc.
8.4 Multiplication by 11 and 9

8.1 SPECIAL NUMBERS

We have already seen the special methods for multiplying numbers near bases. There are many other special multiplication methods in the Vedic system some of which we show here.

8.1a REPEATING NUMBERS

Some multiplications are particularly easy. This type comes under the Sutra *By Mere observation.*

1 **23 × 101 = 2323**.

To multiply 23 by 101 we need 23 hundreds and 23 ones, which gives 2323.

The effect of multiplying any 2-figure by 101 is simply to make it repeat itself.

2 Similarly **69 × 101 = 6969**.

3 And **473 × 1001 = 473473**.

Here we have a 3-figure number multiplied by 1001 which makes the 3-figure number repeat itself.

4 **47 × 1001 = 47047**.

Here, because we want to multiply by 1001, we can think of 47 as 047.
So we get 047047, or just 47047.

5 **123 × 101 = 12423**.

Here we have 12300 + 123 so the 1 has to be carried over.

6 **28 × 10101 = 282828**.

Practice A Find:

a 46 × 101	**b** 246 × 1001	**c** 321 × 1001	**d** 439 × 1001
e 3456 × 10001	**f** 53 × 10101	**g** 74 × 1001	**h** 73 × 101
i 29 × 1010101	**j** 277 × 101	**k** 521 × 101	**l** 616 × 101

a	**4646**	b	**246246**	c	**321321**	d	**439439**
e	**34563456**	f	**535353**	g	**74074**	h	**7373**
i	**29292929**	j	**27977**	k	**52621**	l	**62216**

8.1b *PROPORTIONATELY*

7 **43 × 201 = 8643**.

Here we bring in the *Proportionately* formula: because we want to multiply by 201 rather than 101 we must put twice 43 (which is 86) then 43.

8 **31 × 10203 = 316293**. We have 31×1, 31×2, 31×3.

Practice B Find:

a 54 × 201	**b** 32 × 102	**c** 333 × 1003	**d** 41 × 10201	**e** 33 × 30201
f 17 × 20102	**g** 13 × 105	**h** 234 × 2001	**i** 234 × 1003	**j** 43 × 203

a	**10854**	b	**3264**	c	**333999**	d	**418241**	e	**996633**
f	**341734**	g	**1365**	h	**468234**	i	**234702**	j	**8729**

8.1c DISGUISES

Now it is possible for a sum to be of the above type without it being obvious: it may be disguised.

If we know the factors of some of these special numbers (like 1001, 203 etc.) we can make some sums very easy.

Suppose for example you know that **3 × 67 = 201**.

9 **93 × 67 = 6231**.

Since 3 × 67 = 201,
therefore 93 × 67 = 31 × **3** × **67**
= 31 × **201**
= 6231

In other words, we recognise that one of the special numbers (201 in this case) is contained in the sum (as 3 × 67 here).

Now suppose we know that **3 × 37 = 111.**

10 **24 × 37 = 888**.

We know that 3 × 37 = 111, which is a number very easy to multiply.
So 24 × 37 = 8 × 3 × 37
= 8 × 111
= 888.

Another useful result is 19 × 21 = 399 = $40\bar{1}$.

11 **38 × 63 = 2394**.

38 × 63 = 2×**19** × 3×**21**
= 6 × **19**×**21**
= 6 × $40\bar{1}$
= $240\bar{6}$
= **2394**.

If we know the factors of these special numbers we can make good use of them when they come up in a sum, and they arise quite frequently.

"There, however, are several cases which really belong to this type but come under various kinds of disguises (thin, thick or ultra-thick)!"
From "Vedic Mathematics", Page 111.

Below is a list of a few of these numbers with their factors:

67 × 3 = 201	**17 × 6 = 102**	**11 × 9 = $10\bar{1}$**
43 × 7 = 301	**13 × 8 = 104**	**19 × 21 = $40\bar{1}$**
7 × 11 × 13 = 1001	**29 × 7 = 203**	**23 × 13 = $30\bar{1}$**
3 × 37 = 111	**31 × 13 = 403**	**27 × 37 = $100\bar{1}$**

12 **62 × 39 = 2418.**

We see 31 × 13 contained in this sum: 62 × 39 = 2×**31** × 3×**13**
= 2×3 × **31**×**13**
= 6 × **403**
= 2418.

Practice C Use the special numbers to find:

a 29 × 28 **b** 35 × 43 **c** 67 × 93 **d** 86 × 63

e 77 × 43 **f** 26 × 77 **g** 34 × 72 **h** 57 × 21

i 58 × 63 **j** 26 × 23 **k** 134 × 36 **l** 56 × 29

m 93 × 65 **n** 54 × 74 **o** 39 × 64 **p** 51 × 42

a	812	b	1505	c	6231	d	5418
e	3311	f	2002	g	2448	h	1197
i	3654	j	598	k	4824	l	1624
m	6045	n	3996	o	2496	p	2142

These special numbers can also be useful in finding recurring decimals (see end of Section 5.4).

8.2 USING THE AVERAGE

Here we look at a neat and easy way of multiplying numbers by using their average.
This comes under the formula *Specific General*.

13 Suppose we want to know **29 × 31**.

Since the average of 29 and 31 is 30, we might think that 29 × 31 is 30 × 30, or close to it.
In fact 29 × 31 = **899**
and this is just 1 below 900.

14 Now consider **28 × 32**. Again 30 is their average. 28 × 32 = **896** and this is 4 below 900.

15 For **27 × 33** whose average is also 30: 27 × 33 = **891**, which is 9 below 900.

You may have noticed the rule here:

> square the average
> and subtract the square of the difference of either number from the average.

16 So **26 × 34** = $30^2 - 4^2$ = 900 – 16 = **884**.

And **58 × 62** = $60^2 - 2^2$ = 3600 – 4 = **3596**.

Similarly **94 × 106** = $100^2 - 6^2$ = 10,000 – 36 = **9964**.

And **37 × 33** = $35^2 - 2^2$ = 1225 – 4 = **1221**.

PROOF

$\mathbf{(a + b)(a - b) = a^2 - b^2}$, where a is the average and b the difference of each number from the average. So (a + b) is the higher number and (a – b) is the lower number.

A geometrical explanation for Example 15 is shown below.

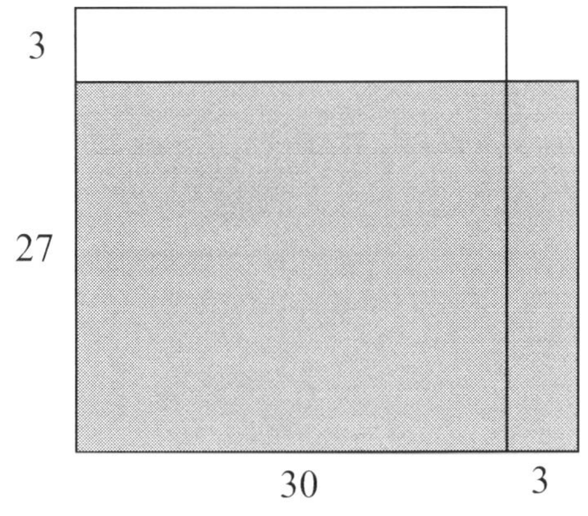

The shaded rectangle is 27 by 33 and its area is 27 ×33.
The superimposed shape is a 30 by 30 square.
This shows that the square whose area is 30^2 is larger than the required rectangle by 3^2 units, as the top rectangle is 3 × 30 and the right-hand rectangle is 27 × 3, a difference of 3 ×3.

This method is available for the product of any two numbers. Even if the average is not a very attractive number this method is still often better than multiplying the numbers. For example, for 67×69 it is easier to find $68^2 - 1$ than to multiply the original numbers together.

✎ **Practice D** Find:

a 49×51	**b** 17×23	**c** 57×63	**d** 64×66	**e** 85×65
f 55×95	**g** 33×47	**h** 91×99	**i** 44×48	**j** 43×47
k 74×86	**l** 98×102	**m** 62×38	**n** 48×72	**o** 73×93
p 196×204	**q** 346×354			

---

a **2499**	b **391**	c **3591**	d **4224**	e **5525**
f **5225**	g **1551**	h **9009**	i **2112**	j **2021**
k **6364**	l **9996**	m **2356**	n **3456**	o **6789**
p **39984**	q **122484**			

8.3 MULTIPLICATION BY NINES

The Vedic formula *By One Less Than the One Before*, which is the converse of the formula of Lesson 4, comes in here in combination with *All From 9 and the Last From 10*.

17 **763 × 999 = 762/237**.

The number being multiplied by 9's is first reduced by 1: 763–1 = **762**, then *All From 9 and the Last From 10* is applied to 763 to get **237**.

18 **1867 × 99999 = 1866/98133**.

Here, as 1867 has 4 figures, and 99999 has 5 figures, we suppose 1867 to be 01867. This is reduced by 1 to give **1866**, and applying *All From 9....* to 01867 gives **98133**.

This is easily understood by considering $763 \times 999 = 763 \times 100\bar{1} = 763\overline{763} = 762/237$.

Practice E

a 89 × 99 **b** 82 × 99 **c** 19 × 99 **d** 45 × 99

e 778 × 999 **f** 79 × 999 **g** 124 × 9999 **h** 8989 × 99999

---

a	8811	b	8118	c	1881	d	4455
e	777222	f	78921	g	1239876	h	898891011

8.4 MULTIPLICATION BY 11 AND 9

The 11 times table is easy to remember and multiplying longer numbers by 11 is also easy.

19 Find **a 52 × 11** and **b 57 × 11**.

a To multiply a 2-figure number, like 52, by 11 we write down the number being multiplied, but put the total of the figures between the two figures: 572.
So 52 × 11 = **572**. Between the 5 and 2 we put 7, which is 5+2.

b For 57 × 11 there is a carry because 5+7=12 which is a 2-figure number.
So we get $5_12 7$ = **627** the 1 in the 12 is carried over to the 5 to give 6.

This is easy to understand because if we want, say, 52×11 we want eleven 52's.
This means we want ten 52's and one 52 or 520 + 52:

$$\begin{array}{r} 5\,2\,0 \\ \underline{5\,2} + \\ \underline{5\,7\,2} \end{array}$$

Note how the 2 and the 5 get added in the middle column.

Practice F Multiply the following by 11:

a 23 **b** 61 **c** 44 **d** 16 **e** 36 **f** 50 **g** 76

h 88 **i** 73 **j** 65 **k** 75 **l** 99

m Is 473 divisible by 11?

---

a	253	b	671	c	484	d	176	e	396	f	550	g	836
h	968	i	803	j	715	k	825	l	1089				

m yes, as 4+3=7

20 Find **234 × 11**.

3-figure numbers like this are also easy to multiply by 11.
234 × 11 = **2574** the first and last figures are the same as in 234.
For the 5 in the answer we add up 2 and 3 and for the 7 we add up 3 and 4.

21 Find **777 × 11**.

The method above gives: $7_{1}4_{1}47$ = **8547**. We simply carry the 1's over.

✎ **Practice G** Multiply by 11:

a 345	**b** 444	**c** 135	**d** 531
e 888	**f** 372	**g** 629	

---

a **3795**	b **4884**	c **1485**	d **5841**
e **9768**	f **4092**	g **6919**	

This method can be extended for multiplying numbers with four or more digits.
For example: 2345 × 11 = 25795.

This can be extended further to products like **234 × 111 = 25974**,
in which we get the digits 2, 5, 9, 7, 4 by taking the first figure **2**, of 234;
then adding the first two figures: 2+3 = **5**;
then adding all three figures: 2+3+4 = **9**;
then adding the last two figures: 3+4 = **7**;
then taking the last figure: **4**.

Carries are dealt with in the usual way.

22 **2536 × 9 = 22824**.

$$2536 \times 9 = 23\bar{2}3\bar{6}$$
$$= 22824$$

We suppose there is a zero at each end of the number, i.e. 025360, and starting with 2 subtract from each figure in turn the figure before it:

$2 - 0 = \mathbf{2}$, $5 - 2 = \mathbf{3}$, $3 - 5 = \bar{\mathbf{2}}$, $6 - 3 = \mathbf{3}$,
$0 - 6 = \bar{\mathbf{6}}$. These are the figures of the answer which then convert to 22824.

The explanation is simply that $9 = 1\bar{1}$ and the moving multiplier method then gives this result.
This can also be done from right to left.

Practice H Multiply by 9:

a 345 **b** 49 **c** 94 **d** 543 **e** 88 **f** 3272

---

a $311\bar{5} = 3105$ b $45\bar{9} = 441$ c $9\bar{5}\bar{4} = 846$ d $5\overline{113} = 4887$ e $80\bar{8} = 792$ f $3\bar{1}5\overline{52} = 29448$

8.5 PERCENTAGES

8.5a INCREASING

Multiplying by 1.1 is equivalent to increasing a number by 10% and multiplying by 1.01 to increasing by 1%, so the methods of multiplying by 11, 101 etc. in this chapter can be very useful.

> We increase a number by 10% by multiplying it by 1.1.

23 **Increase 32 by 10%.**

$32 \times 1.1 =$ **35.2**.

> We increase a number by 1%, 2%, 3% . . .
> by multiplying it by 1.01, 1.02, 1.03 . . .

24 **Increase 43 by 2%.**

Increasing by 2% is the same as multiplying by 1.02 (see Section 8.1b).
And we multiply by 1.02 by multiplying by 102 and placing the decimal point 2 places to the left.
So $43 \times 1.02 =$ **43.86**.

Practice I

Increase by 10%: **a** 24 **b** 17 **c** 74 **d** 232 **e** 258

Increase: **f** 23 by 3% **g** 41 by 2% **h** 88 by 1% **i** 34 by 2%

j 14 by 4% **k** 8 by 1% **l** 19 by 3% **m** 34 by 3%

n 222 by 2% **o** 123 by 3% **p** 50 by 5% **q** 55 by 6%

a	**26.4**	**b**	**18.7**	**c**	**81.4**	**d**	**255.2**	**e**	**283.8**
f	**23.69**	**g**	**41.82**	**h**	**88.88**	**i**	**34.68**		
j	**14.56**	**k**	**8.08**	**l**	**19.57**	**m**	**35.02**		
n	**226.44**	**o**	**126.69**	**p**	**52.5**	**q**	**58.3**		

8.5b REDUCING

To reduce by 1%, 2%, 3% . . .
we multiply by 0.99, 0.98, 0.97 . . .

That is, we multiply by $1.0\bar{1}$, $1.0\bar{2}$, $1.0\bar{3}$ . . .

25 **Reduce 56 by 1%**.

We find 56 × 0.99 = $56 \times 1.0\bar{1} = 56.\overline{56}$ = **55.44**.

26 **Reduce 56 by 3%**.

We find 56 × 0.97 and as $56 \times 1.0\bar{3} = 56_{1}\overline{68}$ = 5432
therefore 56 × 0.97 = **54.32**.

✎ **Practice J** Reduce:

a 33 by 1% **b** 77 by 1% **c** 43 by 2% **d** 78 by 2%

e 32 by 3% **f** 32 by 5% **g** 58 by 3% **h** 47 by 4%

i 11 by 9% **j** 28 by 22%

a	**32.67**	**b**	**76.23**	**c**	**42.14**	**d**	**76.44**
e	**31.04**	**f**	**30.4**	**g**	**56.26**	**h**	**45.12**
i	**10.01**	**j**	**21.84**				

LESSON 9
TRIPLES

SUMMARY

9.1 **Definitions** – of a triple, triple angle etc.
9.2 **Triples for 45°, 30° and 60°**
9.3 **Triple Addition** – adding two triples together.
9.4 **Double Angle** – getting a triple with twice the angle of a given triple.
9.5 **Variations of 3,4,5** – one triple can specify eight directions.
9.6 **Quadrant Angles** – special triples for these angles.
9.7 **Rotations** – using triples to rotate points.

9.1 DEFINITIONS

A **triple** is a set of three numbers in which the sum of the squares of the first two numbers is equal to the square of the third number.

One example of a triple is **3, 4, 5** because $3^2 + 4^2 = 5^2$.

If a triangle had sides of 3, 4 and 5 units it would be right-angled, so the numbers can represent the lengths of the sides of a right-angled triangle.

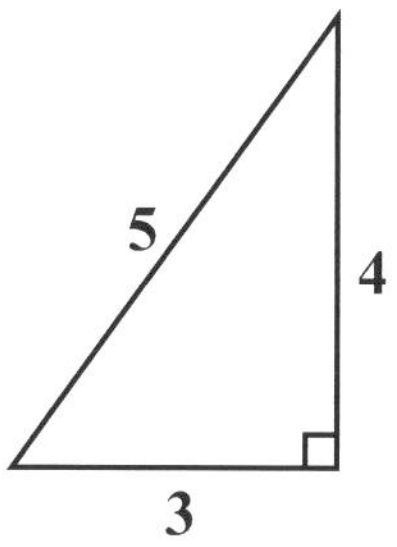

And let us say that the first number in the triple is always the base of the triangle, the second number is always the height and the third is always the hypotenuse.

So **4, 3, 5** represents

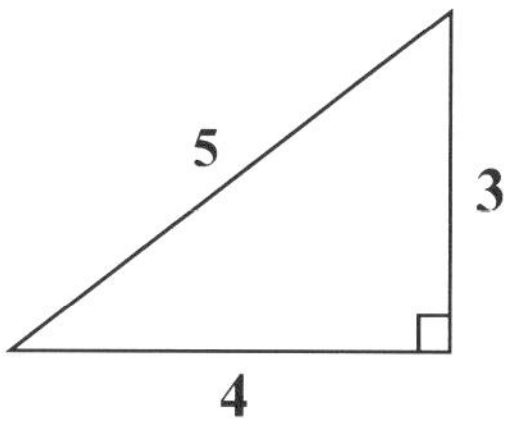

Similarly 2, 1½, 2½; $\sqrt{5}, \sqrt{6}, \sqrt{11}$; $2, \sqrt{5}, 3$ are also triples.

Note: triples that have the same shape are called **equal triples**.
So 36, 15, 39 = 12, 5, 13 = 6, 2½, 6½.

If each of the elements (parts) of a triple are rational numbers then the triple is a **perfect triple**.

So, 2, $\sqrt{5}$,3 is a triple but not a perfect triple as $\sqrt{5}$ is not a rational number.

And 4, 6, 7 is not a perfect triple as it is not even a triple: $4^2 + 6^2$ does not equal 7^2.

The **angle in a triple** is the angle between the **base** and the **hypotenuse**.
So for the **3, 4, 5** triple we can write: **A) 3, 4, 5**

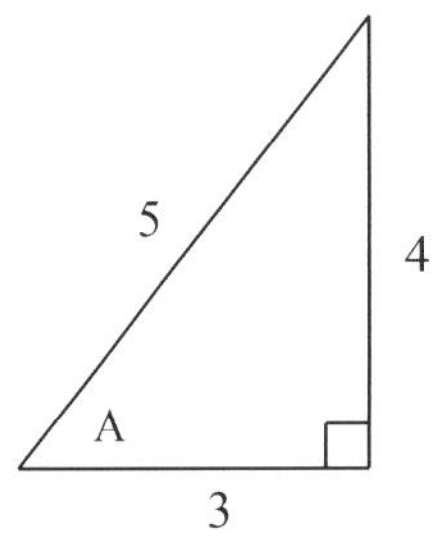

We write the angle first, then a bracket, then the sides of the triple.

Every triangle contains two triples. In this example we also have: **90°–A)4, 3, 5**.

1 Write down the triple which has angle A:

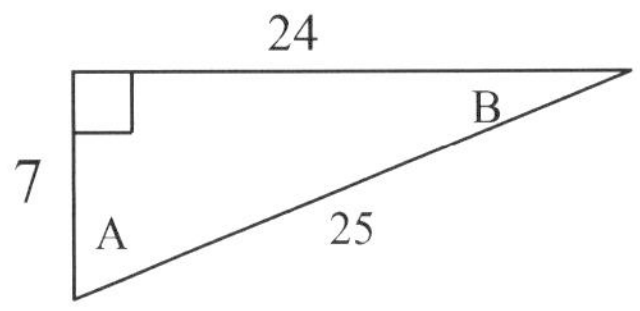

The angle A is between the sides 7 and 25, so 7 must be the base and 25 the hypotenuse.
We therefore put the 7 first: **A)7, 24, 25**.

Similarly if the angle B above was to be described in triple terms it would be: **B)24, 7, 25**.

9.2 TRIPLES FOR 45°, 30° AND 60°

The angles 45°, 30° and 60° are simple angles which frequently occur and can easily be expressed with triples.

If you take a square of side 1 unit and draw a diagonal you will get a triangle with an angle of 45° because the diagonal cuts the right angle in half.

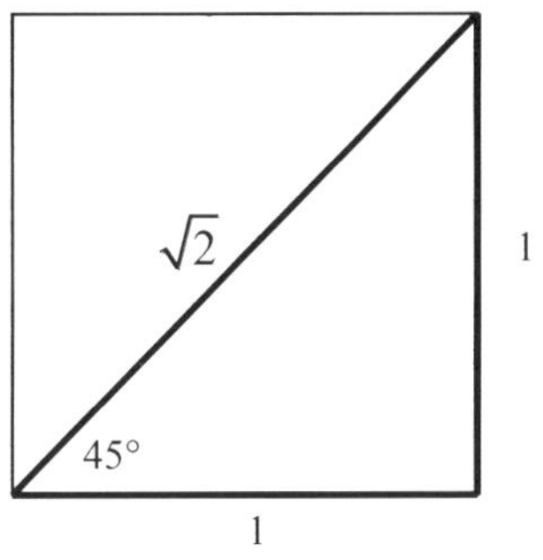

Pythagoras' theorem gives the diagonal length as $\sqrt{2}$ so a triple for 45° will be:

45°) 1, 1, $\sqrt{2}$

Similarly we can take an equilateral triangle of side 2 units and cut it in half:

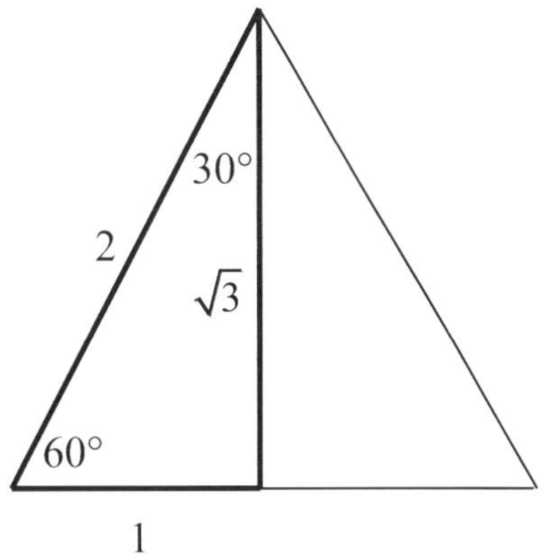

The base of the bold triangle above is 1 unit as the equilateral triangle is cut in half.
For the same reason the 60° angle is cut in half to give 30° at the top.
And Pythagoras' theorem gives the height as $\sqrt{3}$.

❋ From the diagram above write down a triple for 60° and a triple for 30°.
Answers: 60°)1, $\sqrt{3}$, 2, 30°)$\sqrt{3}$, 1, 2.

9.3 TRIPLE ADDITION

Now suppose we have two triples:

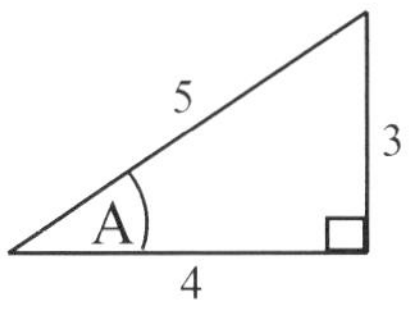

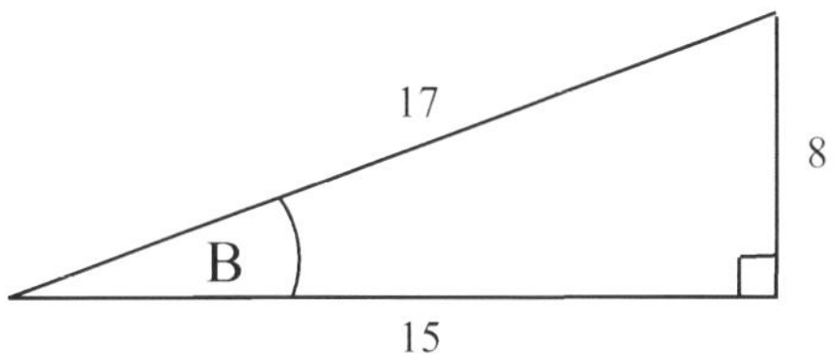

A) 4 3 5 B) 15 8 17

and suppose that we wish to add them in the way shown below, so that the angles are added.

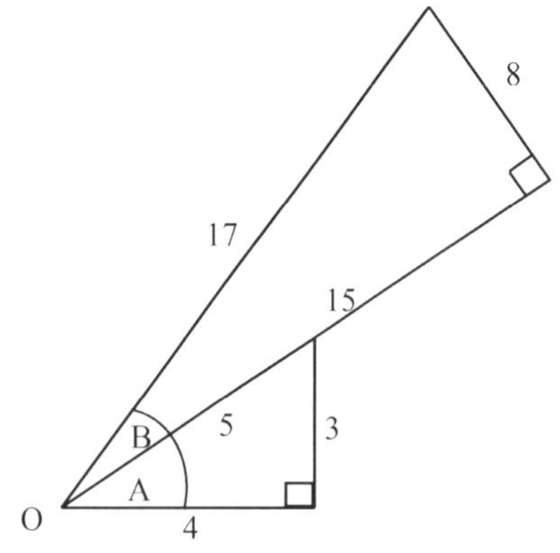

We want the sum of two triples to be itself a triple so we want a triple that contains the angle A+B.

* To skip the following derivation go to * two pages on.

This is shown in bold below.
The triple which contains the angle (A+B) is the triangle OQP below.

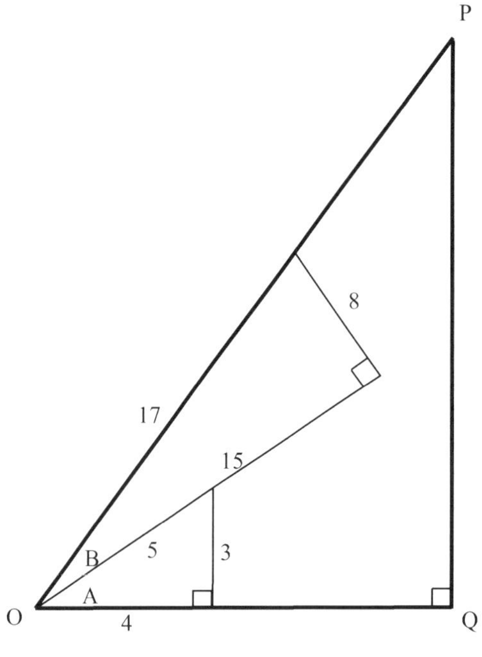

To get the triple for ΔOQP we need the lengths of the sides of ΔOQP.

It will make things simpler if we increase the sides of the 15,8,17 triangle by a factor of 5.
So we use 75,40,85.

This 5 is chosen because 5 is the hypotenuse of the 4,3,5 triangle, and this will make the arithmetic easier.

Of course changing the side of the 15,8,17 triangle will not alter the problem as the angle will still be the same.

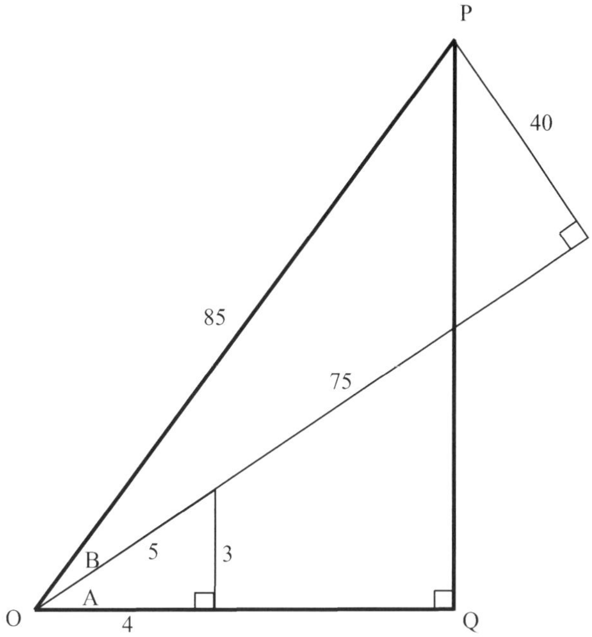

Next we put some more letters on the diagram and add some lines:

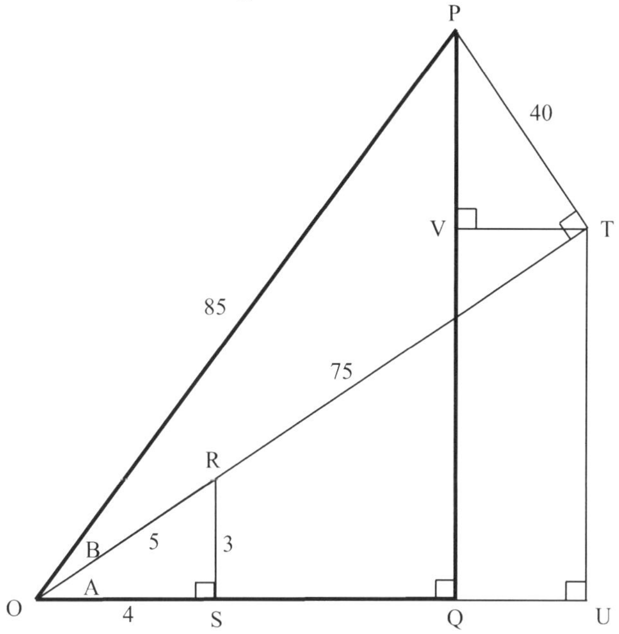

We want the lengths of the sides of ΔOQP and so far we know OP = 85.

ΔOTU is similar to the 4,3,5 triangle. And since its hypotenuse is 75, which is 15 times bigger than the hypotenuse of the 4,3,5 triangle, the height must be TU = **3 × 15 = 45**.

And base OU = **4 × 15 = 60**.

Now ΔPTV is also similar to the 4,3,5 triangle, because $T\hat{P}V = \hat{A}$.

This can be seen by observing that $V\hat{T}O = \hat{A}$ (Z-angles).

And $P\hat{T}V = 90° - A$, which means that $T\hat{P}V = \hat{A}$.

As the hypotenuse of ΔPTV is 40 this makes it bigger than the 4,3,5 triangle by a factor of 8.

So PV = **8 × 4 = 32** and VT = **8 × 3 = 24**.

So now we have:

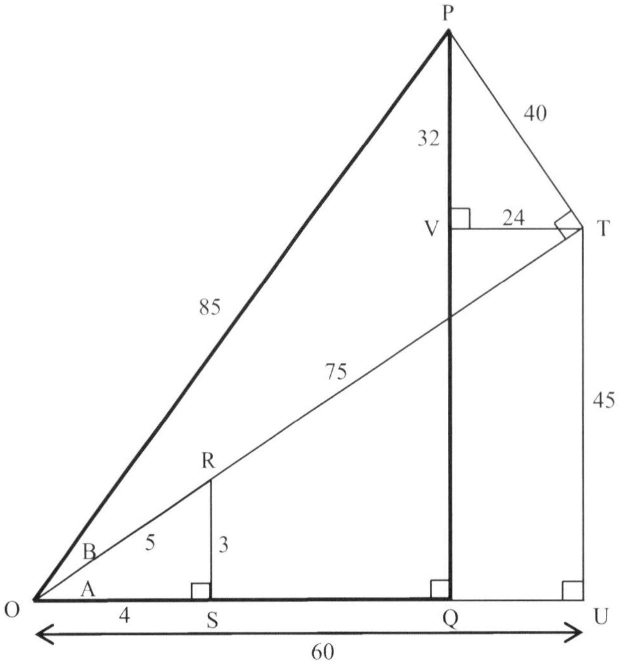

and we can now write the lengths of OQ and QP:

OQ = OU – VT = 60 – 24 = 36, - - - (1)

QP = UT + VP = 45 + 32 = 77. - - - (2)

So a triple for ΔOQP is 36, 77, 85.

Now we ask: how can we get the numbers 33,77,85 directly from the numbers 4,3,5 and 15,8,17?

36 = 60 – 24 (see line (1) above) = 4 × 15 – 3 × 8,
77 = 45 + 32 (see line (2) above) = 3 × 15 + 4 × 8,
and 85 = 5 × 17.

The table below summarises this.

*

A	4	3	5	
B	15	8	17	+
A+B	(4 × 15 – 3 × 8),	(3 × 15 + 4 × 8),	(5 × 17)	
=	36	77	85	

You may see that there is a simple pattern we can use to add the two triples.

That is, we multiply vertically in the first two columns and subtract to get the first element of the triple for (A+B).

Then we cross-multiply in the first two columns and add to get the second element.
And we multiply vertically in the third column to get the third element.

These *Vertical and Crosswise* products are shown diagrammatically below:

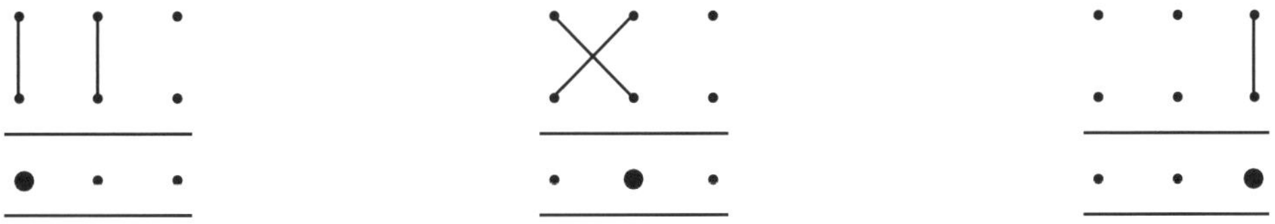

1st element 2nd element 3rd element

Algebraically we can say, for two triples A) x y r and B) X Y R the triple containing the angle (A+B) is given by:

A	x	y	r	
B	X	Y	R	+
A+B	(xX – yY),	(yX + xY),	(rR)	

By replacing the 4,3,5 triangle by x,y,z and the 15,8,17 triangle by X,Y,Z the argument shown leads to the general result above.

2

A	12	5	13	
B	3	4	5	+
A+B	16	63	65	

using the *vertical and crosswise* pattern shown above

Of course we can multiply (or divide) the elements of this triple by any number we like and it will still represent the same angle.
For example, if we want the hypotenuse of the triple 16, 63, 65 to be 13 we can divide through by 5 to get 3.2, 12.6, 13.

3

A	4	3	5	
B	24	7	25	+
A+B	75	100	125	
=	3	4	5	

any common factor can be divided out

4

A	4	3	5	
B	4	3	5	+
A+B	7	24	25	

Practice A Add the following triples:

a 3 4 5	**b** 24 7 25	**c** 40 9 41	**d** 12 5 13	**e** 8 15 17
15 8 17 +	3 4 5 +	4 3 5 +	8 15 17 +	4 3 5 +

a 13, 84, 85 **b** 44, 117, 125 **c** 133, 156, 205 **d** 21, 220, 221 **e** –13, 84, 85

9.4 DOUBLE ANGLE

5

A	12	5	13
A	12	5	13 +
2A	119	120	169

It will be seen from this example and the previous one that the procedure simplifies when finding the triple for a double angle:

A	12 5 13
2A	$(12^2 - 5^2)$, $(2\times12\times5)$, (13^2)

And in general:

A	x y r
2A	$(x^2 - y^2)$, $(2xy)$, (r^2)

6

A	3	4	5
2A	–7	24	25

Here we find that the first element is negative. We can interpret this by considering the addition of two 3,4,5 triangles:

The sum of the two angles is obtuse and the resultant triangle OPQ has its base, OQ, extended in the opposite direction to the base OS.

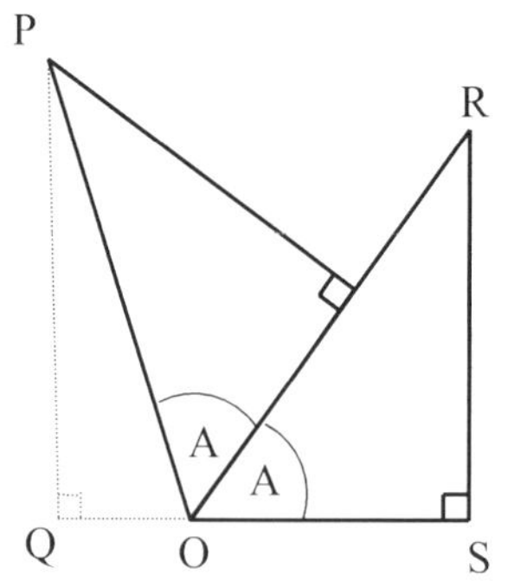

So the triple X) –7, 24, 25 looks like this:

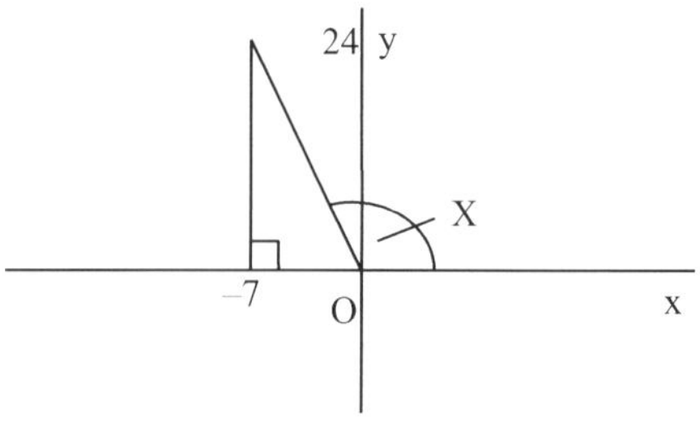

This is rather like the coordinate system we are familiar with, where O is the origin, OS along the x-axis and angles are measured anticlockwise from the positive x-axis.
So if the first element of a triple is negative, and the second positive as we have here, the angle is obtuse (between 90° and 180°).
And conversely, if the angle is obtuse the first two elements are negative and positive respectively.

✎ **Practice B** Find a triple for 2A (note, in **f** and **j** below the dash indicates a number that is not given, but you can find it from the other two numbers):

a A)4, 3, 5	**b** A)15, 8, 17	**c** A) 21, 20, 29	**d** A) 5, 12, 13	**e** A)8, 15,17
f A)9, - , 15	**g** A)1, 2, $\sqrt{5}$	**h** A) $\sqrt{40}$, 3, 7	**i** A)5, $\sqrt{39}$, 8	**j** A)2, - , 5

k Given A)4, 3, 5, find a triple for 3A.

a **7, 24, 25**	b **161, 240, 289**	c **41, 840, 841**	d **–119, 120, 169**	e **–161, 240, 289**
f **–7,24,25**	g **–3,4,5**	h **31,6$\sqrt{40}$,49**	i **-14,10$\sqrt{39}$,64**	j **-17,4$\sqrt{21}$,25**
k **–44,117,125**				

9.5 VARIATIONS OF 3,4,5

One triple, like 3,4,5, can be use to define eight directions in a plane by being put in different positions.

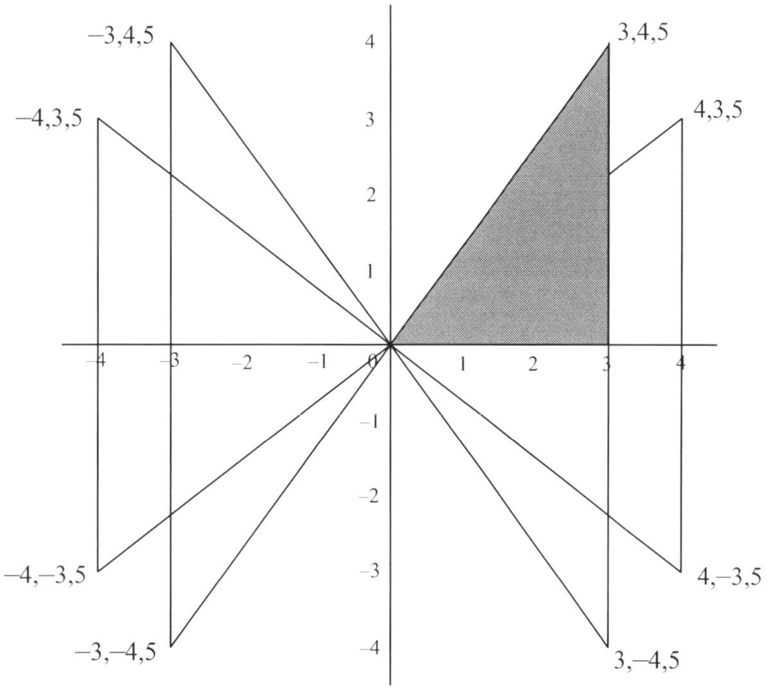

So a triple with only its first element negative, like –3,4,5 is in the second quadrant;
and triple with only its middle element negative, like 3,–4,5 is in the fourth quadrant, because it has a negative height;
and a triple with the first two elements both negative, like –3,–4,5, is in the third quadrant as its base and height are both extended in the negative direction.

So for –5,12,13 the sketch is:

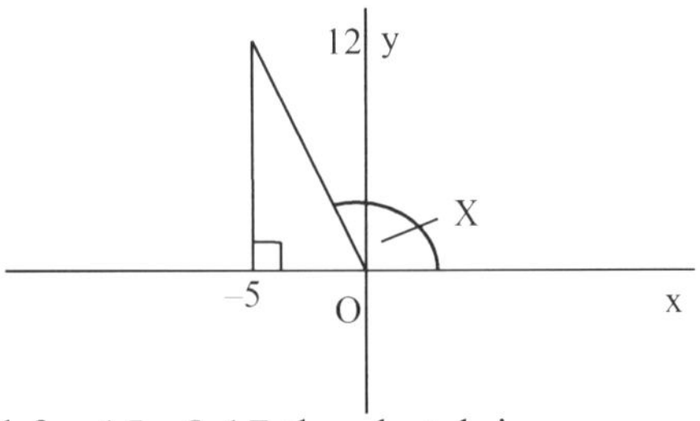

for –12,–5,13 the sketch is:

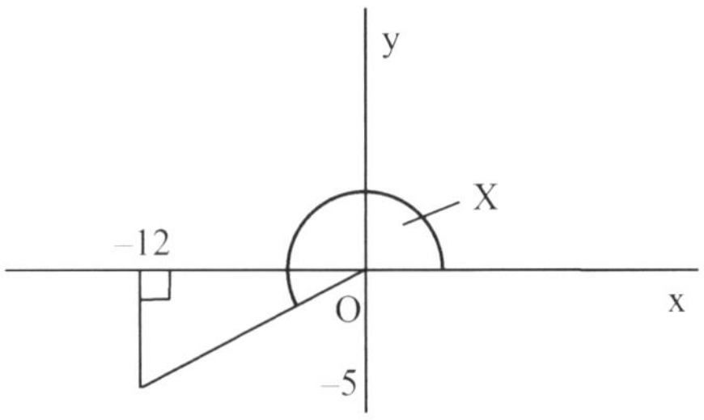

and for 15,–8,17 the sketch is:

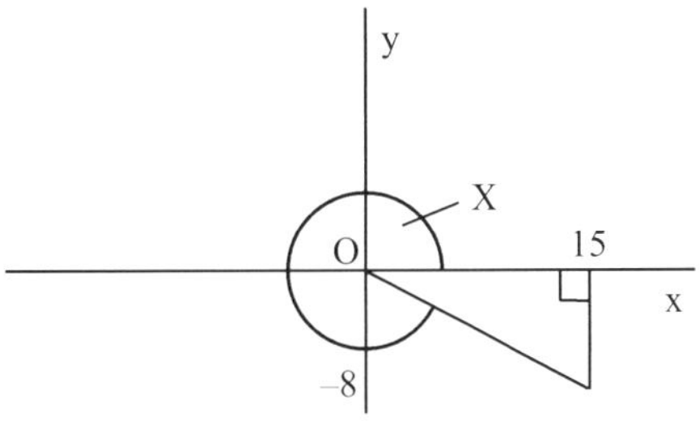

✎ **Practice C** Sketch the following triples:

a 8,–15,17 **b** –7,24,25 **c** $-2,-3,\sqrt{13}$ **d** $-2,\ \sqrt{5},3$

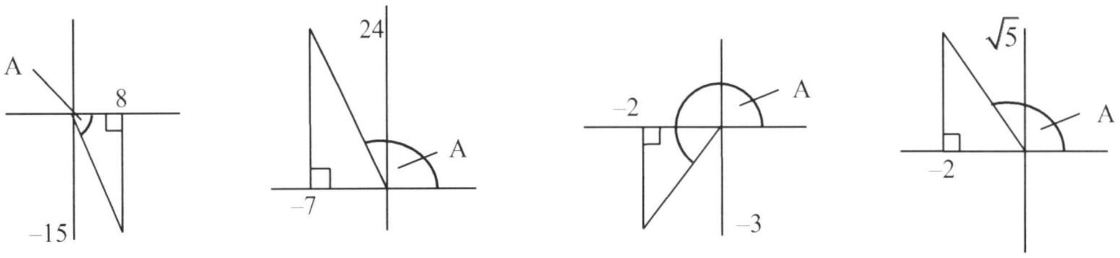

9.6 QUADRANT ANGLES

A quadrant is a quarter of a circle:

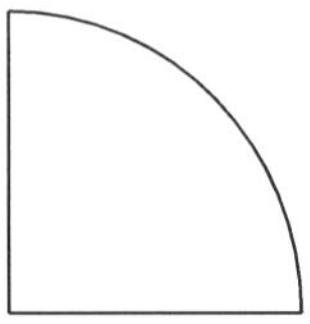

The quadrant angles, **0°**, **90°**, **180°**, **270°** can also be expressed by triples.

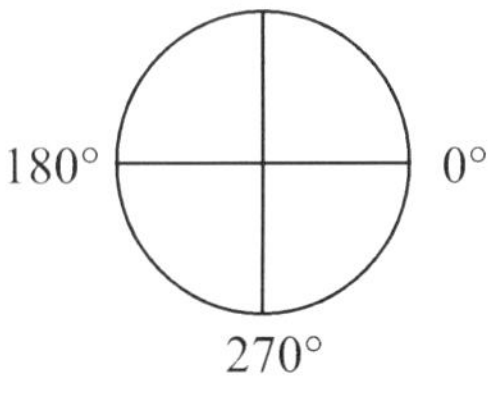

7 Suppose we add the triples **A)4 3 5** and **B)3 4 5**:

A	4	3	5
B	3	4	5 +
A+B	0	25	25
=	0	1	1

We can divide 0,25,25 by 25 to get 0,1,1.

In this case we get a triangle with no base and height and hypotenuse equal.

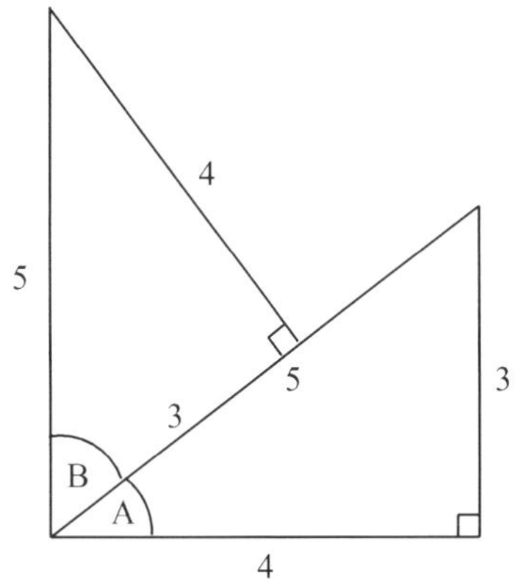

We see that because the triples are complementary A+B = 90°, so that the triple 0, 1, 1 represents an angle of 90°:

90°) 0 1 1

And if 90°) 0 1 1 we can double this triple to get a triple for an angle of 180°:

90°	0	1	1
180°	–1	0	1

Next we can double the 180° triple to get a triple for 360°, which is also a triple for 0°:

180°	–1	0	1
360°	1	0	1

And for 270° we can add 180° and 90°:

180°	–1	0	1
90°	0	1	1
270°	0	–1	1

These results can be summarized:

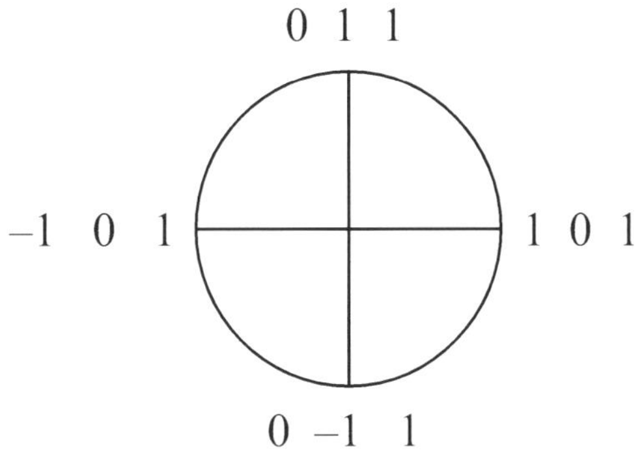

We can now use these triples in combination with others.

8 Given **A) 4 3 5** find a triple for **A+90°**.

A	4	3	5	
90°	0	1	1	+
A+90°	–3	4	5	

9 Given **A) 4 3 5** find a triple for **A+180°**.

A	4	3	5	
180°	–1	0	1	+
A+180°	–4	–3	5	

That is, we reverse the signs of the first two elements when adding 180°.

✐ Practice D

a Given A) 3, 4, 5 find triples for i) 270°+ A,
ii) A + 90°,
iii) 2A + 90°.

b Given A) 4, 3, 5 find A+45°.

c Given A) 5, 12, 13 find a triple for 2A+ 180°.

---

a i) 4, –3,5 ii) –4,3,5 iii) –24,-7,25
b 1,7,5$\sqrt{2}$ c 119,–120,169

10 Find a triple for the angle BAC:

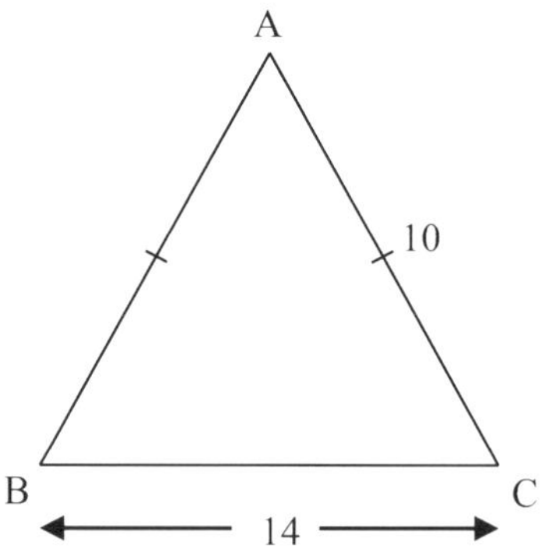

If we call the required angle 2x, then we can get a triple for the half triangle and double it.

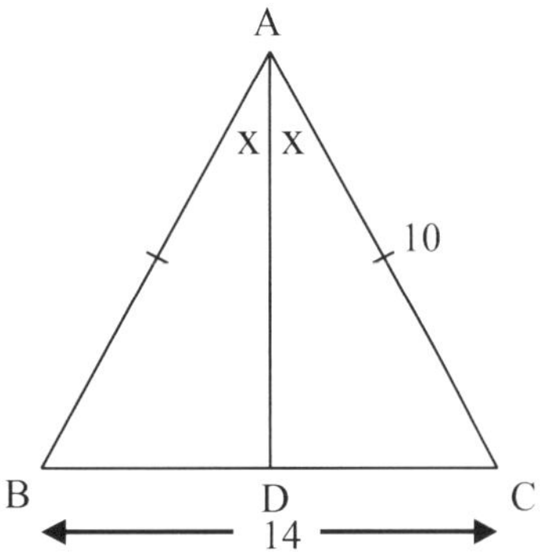

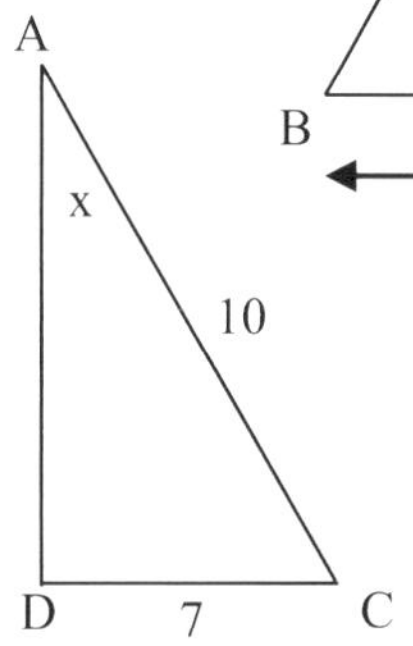

x) - , 7 , 10 = $\sqrt{51}$, 7, 10

$\therefore$ 2x) 2, $14\sqrt{51}$, 100 = 1, $7\sqrt{51}$, 50.

✎ **Practice E** Find a triple for the angle BAC in each of the following triangles.

a

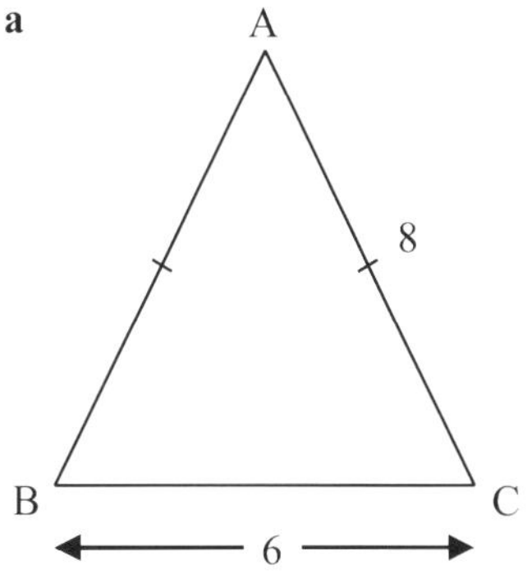

b

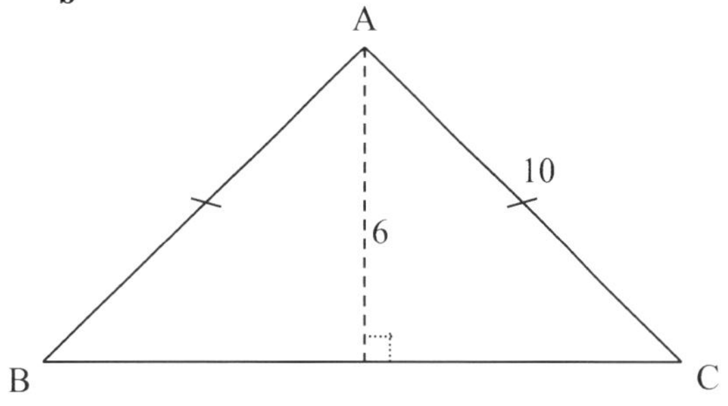

c In a rectangle ABCD side AB = 8cm and side BC = 5cm. If M is the mid point of CD find a triple for angle AMB.

d ABCD is a square and M and N are mid-points.

Find a triple for angle MDN.

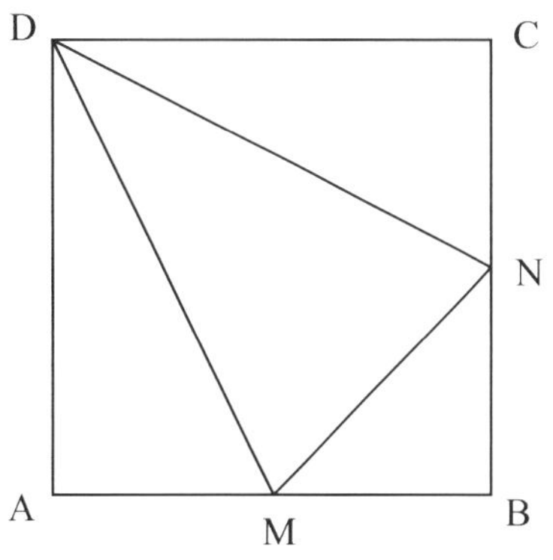

---

a **23,3$\sqrt{55}$,32** **b** **-7,24,25** **c** **9,40,41** **d** **4,3,5**

9.7 ROTATIONS

We will assume all rotations to be anticlockwise unless otherwise stated.

11 **Rotate P(7,3) through an angle of 90° using the origin as centre of rotation**.

We are effectively adding two triangles, one a 7, 3, $\sqrt{58}$ triangle and the other 0,1,1 (triple for 90°).

$$\begin{array}{rrr} 7 & 3 & \sqrt{58} \\ 0 & 1 & 1 \\ \hline -3 & 7 & \sqrt{58} \end{array} \quad +$$

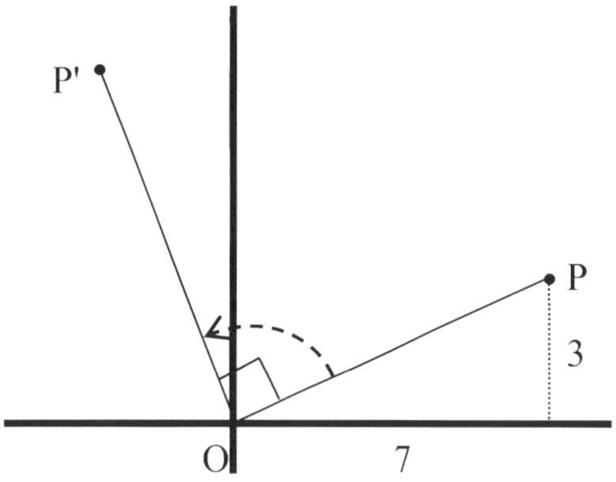

So the position of the rotated point, P', is (–3,7).

Note that since OP=OP' the third column is superfluous.

12 **Rotate the point P(5,2) through an angle A) 4,3,5 about the origin.**

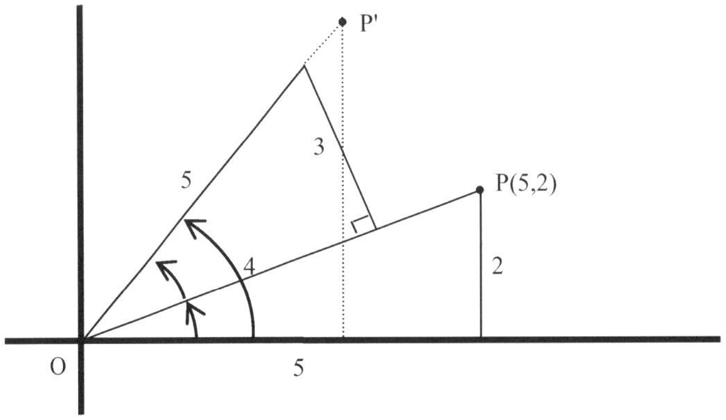

The point (5,2) gives us a right-angled triangle: 5, 2, $\sqrt{29}$.
As the diagram shows, rotating through A) 4,3,5 is equivalent to adding these triples.

$$\begin{array}{ccr} 5 & 2 & \sqrt{29} \\ 4 & 3 & 5 \quad + \\ \hline 14 & 23 & 5\sqrt{29} \\ \hline \end{array}$$

But we want the hypotenuse of the resulting triangle to be equal to OP, which is $\sqrt{29}$.
We therefore divide through by 5 which gives P' $(\frac{14}{5}, \frac{23}{5})$ or P'(2.8, 4.6).

Note that the third column in the calculation above is unnecessary as we consistently divide through by the third element of the rotation triple.

13 **Rotate P(3,5) through A) –4,3,5 about the origin.**

Here we are rotating by an obtuse angle but the method is the same:

$$\begin{array}{ccr} 3 & 5 & - \\ -4 & 3 & 5 \quad + \\ \hline -27 & -11 & - \\ \hline \end{array}$$

$\therefore$ **P'(–5.4,–2.2).**

Practice F Rotate about the origin:

a (3,5) through an angle of 90°

b (–7,2) 180°

c (2,1) by the angle 7,24,25

d (–2,3) by 3,4,5

e (–4,6) by 30°

---

a (–5,3) **b (7,–2)** **c (–0.4,2.2)** **d (–3.6,0.2)** **e** $(-2\sqrt{3}-3, 3\sqrt{3}-2)$

14 **Rotate P(6,5) through A) 3,4,5 about the point (4,1)**

First we transpose the origin to (4,1).
If the origin is at (4,1) then the coordinates of P(6,5) become P'(2,4).

Then we rotate (2,4) by A)3,4,5:

2	4	-	
3	4	5	+
–10	20	-	

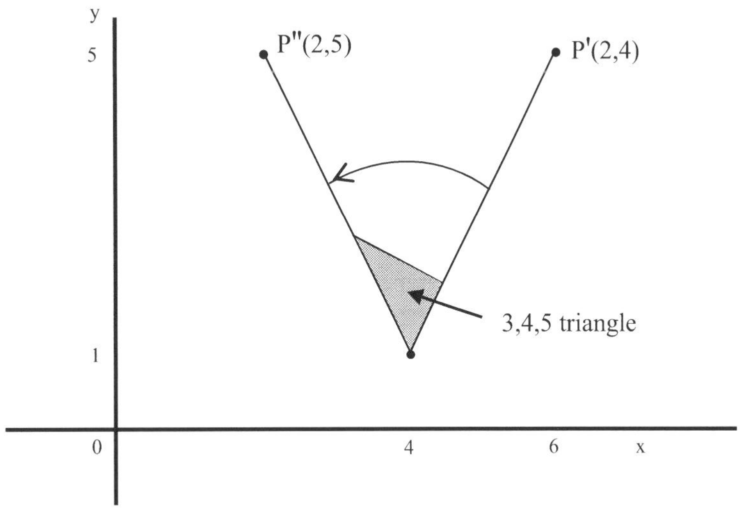

So the rotated point (relative to (4,1) as origin) is (–2,4).
And returning to the original origin by adding (4,1) back on we get **P'(2,5)**.

✎ **Practice G** Rotate:

a (11,4) 90° about the point (3,5).

b (1,–2) 270° about the point (5,5).

c (1,–3) by 4,3,5 about the point (–2,3).

d (–3,1) by 3,–4,5 about (1,2).

---

a (4,13) b (–2,9) c (4,0) d (–2.2,4.6)

We can also rotate points and other objects in 2 and 3-dimensional space so that using triples for rotations can save computer time as we avoid the need to find sines, cosines, square roots etc.

The triple method unifies many areas of mathematics. Many problems that require a special formula of their own can be easily solved by addition or subtraction of triples. Areas of application include coordinate geometry, transformations, trigonometry, simple harmonic motion and astronomy.

LESSON 10
SPECIAL DIVISION

SUMMARY

10.1 Division by Nine – and 8, 11, 12 etc. and algebraic divisions.
10.2 Larger Divisors – dividing by numbers just above or just below a base.
10.3 Proportionately – bringing other sums within the range of 10.2 above.

10.1 DIVISION BY NINE

As we have seen before, the number 9 is special. And there is a very easy way to divide by 9.

10.1a ADDING DIGITS

1 Find **23 ÷ 9**.

The first figure of 23 is the answer: 2.
And we add the figures of 23 to get the remainder: 2 + 3 = 5.
So **23 ÷ 9 = 2 remainder 5**.

It is easy to see why this works because every 10 contains a 9 with 1 left over.
So two tens contains two 9's with 2 left over.
And if 20 contains two 9's remainder 2, then 23 (which is 3 more) contains two 9's remainder 5.

Practice A Divide by 9:

a 51	**b** 34	**c** 17	**d** 44	**e** 60	**f** 71	**g** 46

---

a **5 r6**	b **3 r7**	c **1 r8**	d **4 r8**	e **6 r6**	f **7 r8**	g **4 r10 = 5 r1**

This can be extended to the division of longer numbers.

2 Find **2311 ÷ 9**.

If the sum was written down it would look like this:

```
9)2 3 1  1
  2 5 6 r 7
```

The initial **2** is brought straight down into the answer:

```
9) 2  3  1  1
   ↓
   2
```

This 2 is then added to the 3 in 2311, and **5** is put down:

```
9) 2  3  1  1
    ↗
   2  5
```

This 5 is then added to the 1 in 2311, and **6** is put down again:

```
9) 2  3  1  1
       ↗
   2  5  6
```

This 6 is then added to 1 to give the remainder, **7**:

```
9) 2  3  1  1
          ↗
   2  5  6 rem 7
```

> The first figure of the dividend is the first figure of the answer,
> and each figure in the answer is added to the next figure in the dividend
> to give the next figure of the answer.
> The last number we write down is the remainder.

3 Find **1234 ÷ 9**.

```
9) 1  2  3  4
   1  3  6 r 10
```

In this example we get a remainder of 10, and since this contains another 9 we add 1 to 136 and get **137 remainder 1**.

Practice B Divide the following numbers by 9:

a 212	**b** 3102	**c** 11202	**d** 31	**e** 53
f 203010	**g** 70	**h** 114	**i** 20002	**j** 311101
k 46	**l** 234	**m** 56	**n** 444	**o** 713

a 23 rem 5	**b 344 rem 6**	**c 1244 rem 6**	**d 3 rem 4**	**e 5 rem 8**
f 22556 rem 6	**g 7 rem 7**	**h 12 rem 6**	**i 2222 rem 4**	**j 34566 rem 7**
k 5 rem 1	**l 26**	**m 6 rem 2**	**n 49 rem 3**	**o 79 rem 2**

Example 4 **3172 ÷ 9**.

```
9)3   1   7    2
  3   4  11  r 13
```

Here we find we get an 11 and a 13: the first 1 in the 11 must be carried over to the 4, giving 351, and there is also another 1 in the remainder so we get **352 remainder 4**.

Example 5 Find **21.2 ÷ 9**.

```
9)2   1 . 2  0  0  0
  2 . 3   5  5  5 . . .
```

$= \mathbf{2.3\dot{5}}$

Here we have a decimal point and the answer is given as a decimal, without a remainder. Adding 5 to 0 repeatedly, gives the recurring 5.

✎ **Practice C** Divide the following by 9:

a 6153	**b** 3272	**c** 555	**d** 8252
e 661	**f** 4741	**g** 5747	**h** 2938
i 12345	**j** 75057	**k** 443322	

a **683 rem 6**	b **363 rem 5**	c **61 rem 6**	d **916 rem 8**
e **73 rem 4**	f **526 rem 7**	g **638 rem 5**	h **326 rem 4**
i **1371 rem 6**	j **8339 rem 6**	k **49258**	

10.1b A SHORT CUT

However, to avoid the build-up of 2-figure numbers like 11 and 13, in Example 4 above, we may notice, before we put the 4 down, that the next step will give a 2-figure number and so we put 5 down instead:

```
9)3  1  7  2
  3  5  2 r4
```

Then add 5 to 7 to get 12, but as the 1 has already been carried over we only put the 2 down. Finally, 2+2 = 4.

Example 6 Find **777 ÷ 9**.

```
9)7  7  7
  8  6 r3
```

If we put 7 for the first figure we get 14 at the next step, so we put 8 instead.
Now 8+7 = 15 and the 1 has already been carried over, but if we put the 5 down we see a 2-figure number coming in the next step, so we put 6 down instead.
Then 6+7 = 13 and the 1 has been carried over, so just put down the 3.

Practice D Divide the following by 9, using this short-cut:

a 6153	**b** 3272	**c** 555	**d** 8252
e 661	**f** 4741	**g** 5747	**h** 2938
i 12345	**j** 75057	**k** 443322	**l** 1918161

a	**683 rem 6**	b	**363 rem 5**	c	**61 rem 6**	d	**916 rem 8**
e	**73 rem 4**	f	**526 rem 7**	g	**638 rem 5**	h	**326 rem 4**
i	**1371 rem 6**	j	**8339 rem 6**	k	**49258**	l	**213129**

10.1c DIVIDING BY 8

7 Dividing by 8: **111÷ 8**.

```
8) 1 1|1
   1 3|7
```

When dividing by 8 we can write 2 below it (not shown here), as this is its deficiency from 10. Then instead of putting each number in the answer into the next column, as before, we first double it.

That is: bring down **1**.
Twice that 1 plus 1 = **3**
Twice that 3 plus 1 = **7**.

8 **3411 ÷ 8.**

```
8) 3 4 1|1
   4 2 6|3
```

As in Example 6 we anticipate that the second column will be in excess of 9, and put down 4 initially instead of 3.
And similarly in the third column we put an extra unit.

10.1d ALGEBRAIC DIVISION

9 $\mathbf{(2x^2 + 5x + 7) \div (x - 1)}$.

$$\begin{array}{r|l} x - 1\ ) & 2x^2 + 5x + 8 \\ & \quad 2x \ + \ 7 \text{ rem } 15 \end{array}$$

x–1 is just like number 9 as it is 1 below x, just as 9 is 1 below 10.
This means that the same method can be used in this division as for division by 9.

We bring the first coefficient, 2, down into the answer. (Or dividing *the first by the first*, we get $2x^2 \div x = 2x$). Then we add this 2 to the 5 in the next column and put down 7. Finally adding this 7 to the 8 we get 15 as the remainder.

10 **$(2x^2 + 5x + 8) \div (x - 2)$.**

Here we are dividing by $x - 2$, which is just like dividing by 8.
That is, we double the answer digit before adding it in the next column:

$$\begin{array}{l} x - 2\ )\underline{2x^2 + 5x + 8} \\ \qquad\quad 2x \ + \ 9 \text{ rem } 26 \end{array}$$

This illustrates something we will see over and over again in the Vedic system: that an arithmetic process slides smoothly and perfectly into an algebraic process. This is rarely seen in the conventional system. We can extend this in other directions. For example:

11 **$(2x^2 + 5x + 8) \div (x + 1)$.**

Here we have $(x + 1)$ instead of $(x - 1)$. This means we subtract the last answer coefficient in the next column instead of adding it.

$$\begin{array}{l} x + 1\ )\underline{\ 2x^2 + 5x + 8} \\ \qquad\quad\ 2x \ + 3 \text{ rem } 5 \end{array}$$

Bring down the 2x, then take this 2 from 5 in the next column and put down 3.
Then 3 from 8 is 5 for the remainder.

Practice E

a $101 \div 8$ **b** $1101 \div 8$ **c** $2121 \div 8$ **d** $11111 \div 8$ **e** $132 \div 8$

f $2x^2 + 3x + 4 \div x - 1$ **g** $2x^3 + 3x^2 - 4x + 4 \div x - 1$ **h** $2x^2 + 3x + 4 \div x - 2$

i $2x^2 + 3x + 4 \div x + 1$ **j** $x^3 - 3x + 1 \div x + 3$ *

a	12 rem 5	b	137 rem 5	c	265 rem 1	d	1388 rem 7	e	16 rem 4
f	$2x + 5$ rem 9	g	$2x^2 + 5x + 1$ rem 5	h	$2x + 7$ rem 18				
i	$2x + 1$ rem 3	j	$x^2 - 3x + 6$ rem -17						

* You can insert $0x^2$ here.

This method can be further extended to deal with divisors like $2x + 3$.

10.1e DIVIDING BY 11, 12 etc.

From Example 11 we get the clue as to how to deal with division by 11, 12, 13 etc: we subtract the last answer digit at each step, instead of adding it.

12 **3411 ÷ 11 = 310 rem 1 or** $\mathbf{310.\dot{0}\dot{9}}$.

$$\begin{array}{r} 11\overline{)\,3\ \ 4\ \ 1\ \ 1} \\ \swarrow\ \swarrow\ \swarrow \\ \underline{3\ \ 1\ \ 0\ \text{r}\ 1} \end{array}$$

We bring down the initial **3**.
Then 4 – 3 = **1**.
1 – 1 = **0**.
1 – 0 = **1**.

To decimalise the remainder we just continue the subtractions:

$$\begin{array}{r} 11\overline{)\,3\ \ 4\ \ 1\ \ 1\,.\,0\ \ 0\ \ 0} \\ \swarrow\ \swarrow\ \swarrow\ \swarrow\ \swarrow\ \swarrow \\ \underline{3\ \ 1\ \ 0\,.\,1\ \ \bar{1}\ \ 1\ \ \bar{1}\ldots} \end{array} = \mathbf{310.0909\ldots..}$$

13 **523 ÷ 11 = 47 rem 6**.

$$\begin{array}{r} 11\overline{)\,5\ \ 2\ \ 3} \\ \underline{5\ \ \bar{3}\ \text{r}\ 6} \end{array} = \mathbf{47\ rem\ 6}$$

We get 2 – 5 = $\bar{3}$ here. Then $3-\bar{3}=6$.

14 **3411 ÷ 12 = 284 rem 3**.

$$\begin{array}{r} 12\overline{)\,3\ \ 4\ \ 1\ \ 1} \\ \underline{3\ \ \bar{2}\ \ 5\ \text{r}\ \bar{9}} \end{array} = 285 \text{ rem } \bar{9} = \mathbf{284\ rem\ 3}$$

To divide by 12 we subtract double the last answer digit.
So after bringing down the **3** we have 4 – 2×3 = $\mathbf{\bar{2}}$.
Then 1 – 2× $\bar{2}$ = **5**.
And 1 – 2×5 = $\mathbf{\bar{9}}$.

Since the 285 is 285 twelves we take one of these twelves (leaving 284 of them) to add to the $\bar{9}$ to give a remainder of 3.

Practice F

a 345 ÷ 11 **b** 543 ÷ 11 **c** 20304 ÷ 11 **d** 81726 ÷ 11

e 1489 ÷ 12 **f** 333 ÷ 12

---

a **31 r4** **b** **49 r4** **c** **1845 r9** **d** **$8\bar{6}3\bar{1}$ r7 = 7429 r7**
e **124 r1** **f** **27 r9**

15 **83 ÷ 19 = 4 rem 7**.

$$19\overline{)\,8\quad 3\,}$$
$$\mathbf{4\ rem\ 7}$$

The number of 19's in 83 will be the number of 20's in 80, so we can just divide 8 by 2 and put down **4**. Since every 20 in 80 will have one 19 and one remainder, this answer, 4, is also the remainder after dividing 80 by 19. Then with the 3 the full remainder is **7**.

In other words we divide similarly to division by 9, except that we divide by 2 at each step. This explains the recurring decimal method of Lesson 4.

16 **151.2 ÷ 19 = 7.957**...

$$19\overline{)\,1\quad 5\quad 1\ .\ 2\,}$$
$${}_{1}7\ .\ 9\,{}_{1}5\,{}_{1}7...$$

Similarly here, we divide 15 by 2, put down 7 and prefix the remainder, 1, as 17 is the remainder on dividing 150 by 19. Then 17 + 1 (in the units column) gives 18 to divide by 2 at the next step.

10.2 LARGER DIVISORS

10.2a DIVISOR JUST BELOW A BASE

The method given above for dividing by numbers near a base can be developed for divisors near larger bases.

17 Suppose we want to **divide 235 by 88** (which is close to 100).

We need to know how many times 88 can be taken from 235 and what the remainder is.

Since every 100 must contain an 88 there are clearly 2 88's in 235.
And the remainder will be 2 12's (because 88 is 12 short of 100) plus the 35 in 235.
So the answer is **2 remainder 59** (24+35=59).

A neat way of doing the division is as follows.

8 8)	**2**	**3**	**5**

We separate the two figures on the right because 88 is close to 100 (which has 2 zeros).

Then since 88 is 12 below 100 we put 12 below 88, as shown below.

8 8)	**2**	**3**	**5**
1 2		2	4
	2	**5**	**9**

We bring down the initial 2 into the answer.

This 2 then multiplies the flagged 12 and the 24 is placed under the 35 as shown.
We then simply add up the last 2 columns.

Note that the deficiency of 88 from 100 is given by the formula *All from 9 and the Last from 10.*
Note also that **the position of the vertical line is always determined by the number of noughts in the base number**: if the base number has 4 noughts then the vertical line goes 4 digits from the right, and so on.

This is easily understood since when we bring the initial 2 down into the answer we are expecting to find two 88's in 235. And as there is one 88 in every hundred and 12 left over, in two hundreds there will be two 88's and two 12's remainder which must be added to the 35 to give 59 as the full remainder.

18 **Divide 31313 by 7887**.

We set the sum out as before:

7 8 8 7)	**3**	**1 3 1 3**

7 8 8 7)	**3**	**1 3 1 3**
2 1 1 3		6 3 3 9
	3	**7 6 5 2**

Applying *All From 9 and the Last From 10* to 7887 gives 2113.
Bring the first figure, 3, down into the answer.

We now multiply this by the flagged 2113 and put 6339 in the middle row. Then adding up the last four columns gives the remainder of 7652.

Practice G Divide the following (do as many mentally as you can):

a 88)121	**b** 76)211	**c** 83)132	**d** 98)333
e 887)1223	**f** 867)1513	**g** 779)2222	**h** 765)3001
i 8907)13103	**j** 7999)12321	**k** 7789)21012	**l** 8888)44344

a	1/33	b	2/59	c	1/49	d	3/39
e	1/336	f	1/646	g	2/664	h	3/706
i	1/4196	j	1/4322	k	2/5434	l	4/8792

Next we consider the case where the answer consists of more than one digit.

19 **1108 ÷ 79 = 13 remainder 81 = 14 remainder 2**.

We set out the sum marking off two figures on the right (as we have a 2-figure divisor) and leave two rows as there are to be two answer figures.

7 9) 1 1	**0 8**

7 9) 1 1	**0 8**	
2 1 2	1	
	6 3	
1 3	**8 1**	= **14** remainder **2**

Bring the first 1 down into the answer.
Multiply the flagged 21 by this 1 and put the answer (21) as shown in the second row.
Adding in the second column we get 3 which we put down and then multiply the 21 by this 3 to get 63, which we place as shown in the third row.

Add up the last two columns, but since the remainder, 81, is greater than the divisor, 79, there is another 79 contained in 81 so there are 14 79's in 1108 with 2 remaining.

20 **1121123 ÷ 8989**.

8 9 8 9) 1 1 2	1 1 2 3
1 0 1 1 1 0	1 1
2	0 2 2
	4 0 4 4
1 2 4	6 4 8 7

The initial 1 comes down into the answer and multiplies the flagged 1011. This is placed as shown in the second row.

Adding in the second column we put 2 down in the answer and then multiply the 1011 by it. Put 2022 in the third row.

Adding in the third column we get 4 which we put down and also multiply by 1011. So we put 4044 in the fourth row and then add up the last four columns to get the remainder.

> Once the vertical line has been drawn in you can see the number of lines of working needed: this is the number of figures to the left of this line (3 figures and therefore 3 lines of working in the example above).

21 Find **1012 ÷ 898** to 4 S.F.

898)	**1**	**0**	**1**	**2**	**.0**	**0**	**0**
1 0 2		1	0	2			
			1	0	2		
				2	0	4	
	1.	**1**	**2**	**7**			

The decimal point in the answer goes at the same place as the vertical line would have done. We bring down the initial 1 and multiply the deficiency figures 1, 0, 2 by this and put them in the next three columns.
Then add up the next column, put down 1, which we multiply by the 1, 0, 2 and put up 1, 0, 2 in the next columns.
Add the third column, put down 2 and multiply this by 1, 0, 2 and put up 2, 0, 4.
This gives 6 in the 4th column, but (mentally) the next step will give 2+7 so the 6 will in fact be 7.
So **1012 ÷ 898 = 1.127** to 4 S.F.

A SIMPLIFICATION

In these examples (and in the ones in the next section) the lines of working can be dispensed with by using the *Vertically and Crosswise* formula. We use the vertical and crosswise products in the flag and answer digits.

In Example 19 we have 21 flagged and the first answer figure is 1:

2 1
1 -

The first vertical product here gives 2×1=2 which is to be added in the second column of 1108 to give 3 as the second answer figure:

2 1
1 3

So now we take the cross-product $2 \times 3 + 1 \times 1 = 7$ and add this to the 0 in 1108 to give 7 as the first remainder figure.
Finally the vertical product on the right in

$$\begin{matrix} \mathbf{2} & \mathbf{1} \\ \mathbf{1} & \mathbf{3} \end{matrix}$$

gives $1 \times 3 = 3$ to be added to the last figure of 1108 which makes 11 and gives the full remainder of $7_11 = 81$.
Similarly longer sums like Example 20 can also be dealt with in this way.

Practice H Divide the following:

a 89)1021 **b** 88)1122 **c** 79)1001 **d** 88)2111 **e** 97)1111

f 888) 10011 **g** 887)11243 **h** 899)21212 **i** 988)30125 **j** 8899)201020

a	11/42	b	12/66	c	12/53	d	23/87	e	11/44
f	11/243	g	12/599	h	23/535	i	30/485	j	22/5242

10.2b DIVISOR JUST ABOVE A BASE

A very similar method, but under the formula *Transpose and Apply* allows us to divide numbers which are close to but above a base number.

22 **1489 ÷ 123 = 12 remainder 13.**

Here we see that 123 is close to the base of 100 so we mark 2 figures off on the right. In fact the method is just as before except that we write the flagged numbers as bar numbers:

$$\begin{array}{lll} 1\,2\,3\,) & 1 \;\; 4 & 8 \;\; 9 \\ \;\;\bar{2}\,\bar{3} & \;\;\;\; \bar{2} & \bar{3} \\ & & \bar{4} \;\; \bar{6} \\ \hline & 1 \;\; 2 & 1 \;\; 3 \end{array}$$

Bring the initial 1 down into the answer.
Multiply this 1 by the flagged $\overline{23}$ and write $\bar{2}$, $\bar{3}$.
Add in the second column and put down 2.
Multiply this 2 by the $\overline{23}$ and put down $\bar{4}$, $\bar{6}$.
Then add up the last two columns.

Practice I Divide the following:

a 123)1377 **b** 131)1481 **c** 121)256 **d** 132)1366

e 1212)13545 **f** 161)1781 **g** 1003)321987 **h** 111)79999

a	11/24	b	11/40	c	2/14	d	10/46
e	11/213	f	11/10	g	321/024	h	720/79

Two other variations, where negative numbers come into the answer or remainder are worth noting.

23 Find **10121 ÷ 113**.

113)	**1**	**0**	**1**	**2**	**1**
$\bar{1}\,\bar{3}$		$\bar{1}$	$\bar{3}$		
			1	3	
				1	3
	1	$\bar{\mathbf{1}}$	$\bar{\mathbf{1}}$	**6**	**4**

When we come to the second column we find we have to bring $\bar{1}$ down into the answer, multiplying this by the flagged $\bar{1}\,\bar{3}$ means we add 13 in the third row (two minuses make a plus).

The answer $1\bar{1}\bar{1}$ we finally arrive at is the same as 100 – 11 which is 89.

24 Find **2211 ÷ 112**.

112)	**2**	**2**	**1**	**1**
$\bar{1}\,\bar{2}$		$\bar{2}$	$\bar{4}$	
			0	0
	2	**0**	$\bar{\mathbf{3}}$	**1**

$= 20$ rem $\overline{29}$ or **19 rem 83**

20 remainder –29 means that 2211 is 29 short of 20 112's.
This means there are only 19 112's in 2211, so **we add 112 to –29** to get 19 remainder 83.

The same simplification is available as in the previous section.

Practice J Divide the following:

a 112)1234 **b** 121)3993 **c** 103)432 **d** 1012)21312

e 122)3333 **f** 123)2584 **g** 113)13696 **h** 1212)137987

i 111)77665 **j** 121)2652 **k** 1231)33033

---

a	**11/02**	b	**33/00**	c	**4/20**	d	**21/060**
e	**27/39**	f	**21/01**	g	**121/23**	h	**113/1031**
i	**699/76**	j	**21/111**	k	**26/1027**		

10.3 PROPORTIONATELY

Of course not all divisors are close to a base number. But the above method can be extended a great deal with a little ingenuity, and the *Proportionately* Sutra.

25 Find **346 ÷ 24**.

Here we could multiply both numbers by 4 to get a divisor close to 100: **1384 ÷ 96**.

$$\begin{array}{rl|ccccc}
9 & 6\,) & 1 & 4 & \bar{2} & 4 & .\,0 \\
{\scriptstyle 0} & {\scriptstyle 4} & & {\scriptstyle 0} & {\scriptstyle 4} & & \\
 & & & & {\scriptstyle 2} & {\scriptstyle \bar{4}} & \\
 & & & & & {\scriptstyle 2} & {\scriptstyle \bar{4}} \\
\hline
 & & 1 & 4 & .\,4 & 2 &
\end{array}$$

We put $14\bar{2}4$ instead of 1384 to avoid the large 8.
At the second stage, where we multiply 0, 4 by 4 and get 0,16, we could put down 1, 6 in the 3rd and 4th columns, but seeing a 4 already in that column 2, $\bar{4}$ is better than 1, 6.

Had we needed 1231 ÷ 24 it would be better to find 1231 ÷ 96 and multiply the answer obtained by 4. This is because the initial 1 in 1231 is to our advantage.

26 **63 ÷ 52 = 1.212 to 4 S.F.**

Here we can double both numbers and find **126 ÷ 104**.

$$\begin{array}{rl|ccccc}
10 & 4\,) & 1 & 2 & 6 & .\;0 & 0 \\
{\scriptstyle 0} & {\scriptstyle \bar{4}} & & {\scriptstyle 0} & {\scriptstyle \bar{4}} & & \\
 & & & & {\scriptstyle \bar{1}} & {\scriptstyle 2} & \\
 & & & & & {\scriptstyle 0} & {\scriptstyle \bar{4}} \\
\hline
 & & 1 & .\,2 & 1 & 2 &
\end{array}$$

Note here that at the second step, where we multiply 0 $\bar{4}$ by 2 to get 0 $\bar{8}$ we can write this as $\bar{1}$ 2 as shown.

Practice K Find to 4 S.F.

a 678 ÷ 49 **b** 678 ÷ 52 **c** 678 ÷ 33 **d** 44.6 ÷ 34 **e** 112.2 ÷ 34

---

a **13.84** b **13.04** c **20.55** d **1.312** e **3.3**

LESSON 11
STRAIGHT DIVISION

SUMMARY

11.1 Single Figure on the Flag – one-line division by 2-figure numbers.
11.2 Short Division Digression – choosing the remainder you want.
11.3 Longer Numbers – dividing numbers of any size.
11.4 Decimalising the Remainder
11.5 Negative Flag Digits – using bar numbers to simplify the work.
11.6 Larger Divisors
11.7 Algebraic Division

11.1 SINGLE FIGURE ON THE FLAG

Straight division is the general division method by which any numbers of any size can be divided in one line. Sri Bharati Krsna Tirthaji, the man who rediscovered the Vedic system, called this "the crowning gem of Vedic Mathematics".
It comes under the *Vertically and Crosswise* Sutra.

1 Suppose we want to **divide 209 by 52**.

We need to know how many 52's there are in 209.
Looking at the first figures we see that since 5 goes into 20 four times we can expect four 52's in 209.

We now take four 52's from 209 to see what is left.
Taking four 50's from 209 leaves 9 and we need to take four 2's away as well.
This leaves a remainder of 1.

We can set the sum out like this:

$$\begin{array}{r|r|l} 2 & 2\ \ 0 & 9 \\ 5 & & {}_{0} \\ \hline & 4 & 1 \end{array}$$

The **divisor**, **52**, is written with the 2 raised up, *On the Flag*, and a vertical line is drawn one figure from the right-hand end to separate the answer, 4, from the remainder, 1.

The steps are:

A. 5 into 20 goes 4 remainder 0, as shown.
B. Answer digit 4 multiplied by the flagged 2 gives 8, and this 8 taken from 9 leaves the remainder of 1, as shown.

2 **Divide 321 by 63.**

We set the sum out:

	3	3 2	1
6			$_2$
		5	6

= 5 remainder 6

6 into 32 goes 5 remainder 2, as shown,
and answer, 5, multiplied by the flagged 3 gives 15, which we take from the 21 to leave the remainder of 6.

What we are doing here is subtracting five 60's from 321, which leaves 21 and then subtracting five 3's from the 21. That means we have subtracted five 63's and 6 is left.

In the following exercise set the sums out as shown above.

Practice A Divide the following:

a 103 ÷ 43 **b** 234 ÷ 54 **c** 74 ÷ 23 **d** 504 ÷ 72

e 444 ÷ 63 **f** 543 ÷ 82 **g** 567 ÷ 93

---

a **2r17** b **4r18** c **3r5** d **7r0**
e **7r3** f **6r51** g **6r9**

11.2 SHORT DIVISION DIGRESSION

Suppose we want to divide 3 into 10.
The answer is clearly 3 remainder 1: 3) 1 0
3 rem 1

But other answers are possible: 3) 1 0 / 2 rem 4 or 3) 1 0 / 1 rem 7 or even 3) 1 0 / 4 rem $\bar{2}$

Since all of these are correct we can select the one which is best for a particular sum.

Practice B Copy each of the following sums and replace the question mark with the correct number:

a 5) 2 1 — 3 rem ?

b 7) 5 1 — 6 rem ?

c 4) 3 0 — 6 rem ?

d 3) 2 2 — ? rem 4

e 5) 4 2 — 6 rem ?

f 6) 3 9 — 4 rem ?

g 5) 2 4 — 5 rem ?

h 7) 2 6 — 4 rem ?

---

a 6 **b** 9 **c** 6 **d** 6

e 12 **f** 15 **g** $\bar{1}$ **h** $\bar{2}$

3 **503 ÷ 72.**

If we proceed as before:

$$\begin{array}{r|cc|l} 2 & 5 & 0 & 3 \\ 7 & & & {}_1 \\ \hline & & 7 & \end{array}$$

We find we have to take 14 from 13, which means the answer is 7 rem $\bar{1}$.

If a negative remainder is not acceptable however we can say that 7 into 50 in the sum above is not 7 rem 1, but 6 remainder 8:

$$\begin{array}{r|cc|l} 2 & 5 & 0 & 3 \\ 7 & & & {}_8 \\ \hline & & 6 & 71 \end{array}$$

Then we find we can take 12 from 83 to get the positive remainder 71.

This reducing of the answer figure by 1 or 2 is sometimes necessary if negative numbers are to be avoided. But it is worth noting that when the answer figure is reduced by 1 the remainder is increased by the first figure of the divisor: so in the answer above the 7 rem 1 is replaced by 6 rem 8: the remainder is increased by 7, the first figure of 72.

Continuing the above example with the first method we would get:

$$\begin{array}{r|cc|l} 2 & 5 & 0 & 3 \\ 7 & & & {}_1 \\ \hline & & 7 & \bar{1} \end{array} = 6 \text{ rem } 71.$$

The 7 we get in the answer represents seven 72's, so we take one of these (leaving 6 of them) and add it to the negative remainder to get $72 + \bar{1} = 71$ for the remainder.

Practice C Divide the following:

a 97 ÷ 28 **b** 184 ÷ 47 **c** 210 ÷ 53 **d** 373 ÷ 63 **e** 353 ÷ 52

f 333 ÷ 44 **g** 267 ÷ 37 **h** 357 ÷ 59 **i** 353 ÷ 59

---

a 3r13 **b** 3r43 **c** 3r51 **d** 5r58 **e** 6r41
f 7r25 **g** 7r8 **h** 6r3 **i** 5r58

11.3 LONGER NUMBERS

4 **17496 ÷ 72.**

The procedure is just the same as before and goes in cycles.

We set the sum out in the usual way:

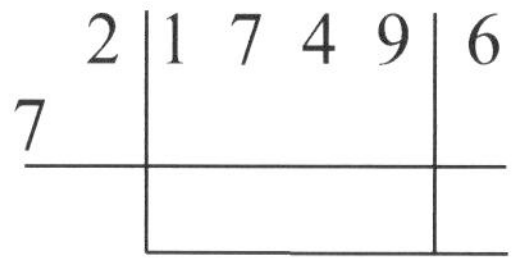

Then we divide 7 into 17 and put down 2 remainder 3.
Note the diagonal of numbers: 2, 3, 4.

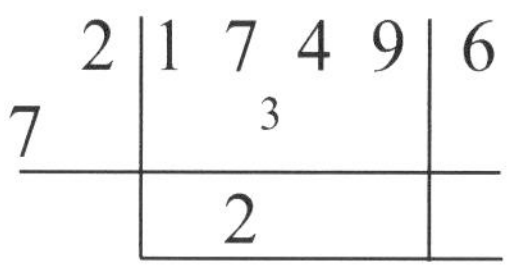

Next we multiply the answer figure by the flag figure: 2×2=4, take this from the 34 to get 30, and then divide by 7 again, to get 4 remainder 2, as shown:

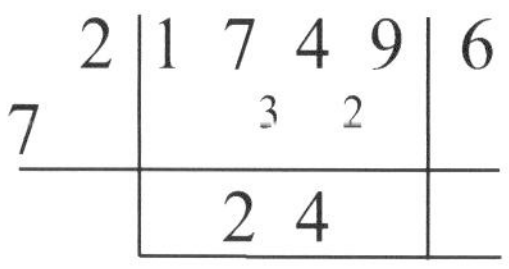

Then we repeat: multiply the last answer figure by the flag to get 8, take this from 29 to get 21, then 7 into 21 goes 3 remainder 0, as shown:

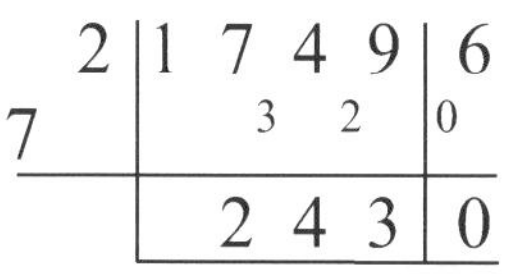

Finally we again multiply the last answer figure by the flag to get 6 and take this from the 6 to get a remainder of 0.

It is important to note that we proceed in cycles as shown in the diagrams above.
Each cycle is completed as each diagonal goes down.

> Each cycle consists of:
> **A.** multiplying the last answer figure by the flag,
> **B.** taking this from the number indicated by the top two figures of the diagonal,
> **C.** dividing the result by the first figure of the divisor and putting down the answer and remainder.

That is (divide), multiply, subtract, divide;
multiply, subtract, divide; . . .

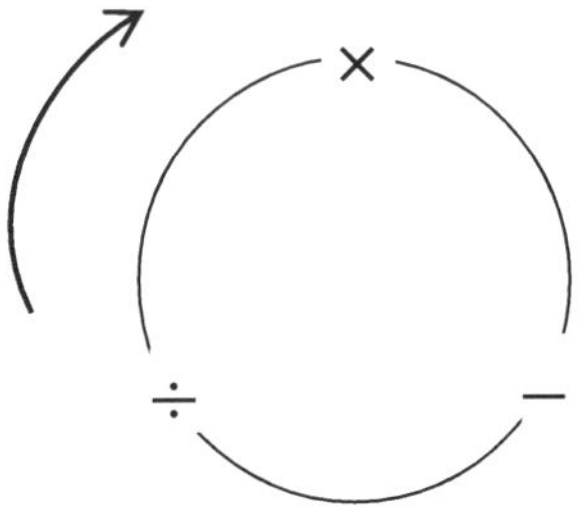

5 **50607 ÷ 123.= 411 rem 54.**

Although the divisor has three digits here dividing by 12 is not a problem and so we can use the same procedure:

3	5	0	6	0	7
12		2	2		5
		4	1	1	54

Practice D Divide the following (the remainder is zero for the first four sums, so you will know if it is correct):

a 19902 ÷ 62 **b** 44749 ÷ 73 **c** 1936 ÷ 88 **d** 4032 ÷ 72

e 4154 ÷ 92 **f** 23824 ÷ 51 **g** 92054 ÷ 63 **h** 142857 ÷ 61

i 12233 ÷ 53 **j** 9018 ÷ 71 **k** 8910 ÷ 72 **l** 23658 ÷ 112

m 40000000 ÷ 61 **n** 14018 ÷ 64 **o** 4712 ÷ 45 **p** 22222 ÷ 76

q 651258 ÷ 82 **r** 301291 ÷ 56 **s** 511717 ÷ 73 **t** 360293 ÷ 46

---

a	**321**	b	**613**	c	**22**	d	**56**
e	**45r14**	f	**467r7**	g	**1461r11**	h	**2341r56**
i	**230r43**	j	**127r1**	k	**123r54**	l	**211r26**
m	**655737r43**	n	**219r2**	o	**104r32**	p	**292r30**
q	**7942r14**	r	**5380r11**	s	**7009r60**	t	**7832r21**

MULTIPLICATION REVERSED

Straight division can also be demonstrated by reversing the vertically and crosswise multiplication method. Given 4032÷72 for example:

$$\begin{array}{r} p\ q \\ \underline{7\ 2} \\ \underline{4\,0\,3\,2} \end{array}$$

We need the values of p and q so that the number pq multiplied by 72 gives 4032.
We see p must be 5 because p multiplied by 7 must account for the 40 in 4032 (or most of it).
And since 5×7=35 there is a remainder of 5:

$$\begin{array}{r} 5\quad q \\ \underline{7\quad 2} \\ \underline{4\ 0\ {}_{5}3\ 2} \end{array}$$

We are left with 532 to be accounted for by the crosswise multiplication and the vertical product on the right. Considering the crosswise part we see we have 5×2=10 and we can take this off the 53 in 532 to leave 43: to be produced by the other part of the crosswise product, 7×q. This tells us that q must be 6 and there is a remainder of 1 from the 53:

$$\begin{array}{r} 5\quad 6 \\ \underline{7\quad 2} \\ \underline{4\ 0\ {}_{5}3\ {}_{1}2} \end{array}$$

The 12 now in the right-hand place is then fully accounted for by the vertical product on the right, so there is no remainder.
All divisions can be done in this way, as a reversal of the multiplication process, and the *on the flag* method in this chapter can be derived from it.

11.4 DECIMALISING THE REMAINDER

We can continue the division when the remainder is reached and give the answer to as many decimal places as required.

6 Find **40342 ÷ 73** to 5 decimal places.

$$\begin{array}{r|ccccccccccc} 3 & 4 & 0 & {}_{5}3 & {}_{3}4 & {}_{5}2 & {}_{4}.0 & {}_{1}0 & {}_{1}0 & {}_{3}0 & {}_{6}0 & {}_{2}0 \\ 7 & & & & & & & & & & & \\ \hline & & 5 & 5 & 2. & 6 & 3 & 0 & 1 & 3 & 7 & \\ \hline \end{array}$$

To give an answer correct to 5 decimal places we should find 6 figures after the point in case we need to round up. So we put a decimal point and six zeros after 40342.

The decimal point in the answer goes where the vertical line went before, one figure to the left of the last figure of the number being divided.
We proceed as usual: multiply by the flag, subtract, divide by 7 for each cycle.

So the answer is **552.63014** to 5 decimal places.

7 Find **23.1 ÷ 83** to 3 decimal places.

The answer is clearly less than 1 because 23 is less than 83.

$$\begin{array}{r|l} 3 & 2\ 3.\ 1\ 0\ 0\ 0 \\ 8 & {\scriptstyle 7\ \ 9\ \ 5\ \ 2} \\ \hline & 0\ .2\ 7\ 8\ 3 \end{array}$$

As before the decimal point goes one figure to the left in the answer, which is **0.278**.

Practice E Find to 2 decimal places:

a 40342 ÷ 73 **b** 371426 ÷ 81 **c** 888 ÷ 61 **d** 17 ÷ 72

e 89179 ÷ 53 **f** 209029 ÷ 85 **g** 22.22 ÷ 36 **h** 104077 ÷ 59

a	**552.63**	b	**4585.51**	c	**14.56**	d	**0.24**
e	**1682.62**	f	**2459.16**	g	**0.62**	h	**1764.02**

11.5 NEGATIVE FLAG DIGITS

When the flag number is large we often need to reduce more frequently. It is possible to avoid these reductions however by using negative flag digits.

8 **97 ÷ 28**.

If we proceed as usual we get:

$$\begin{array}{r|l|l} 8 & 9 & 7 \\ 2 & & {\scriptstyle 3} \\ \hline & 3 & 13 \end{array}$$

We have to reduce the answer digit from 4 to 3 so that the remainder is big enough.

These reductions occur more frequently when the flag number is large (8 here).
This can be avoided however by re-writing 28 as $3\bar{2}$.

$$\begin{array}{r|l|l} \bar{2} & 9 & 7 \\ 3 & & {\scriptstyle 0} \\ \hline & 3 & 13 \end{array}$$

3 into 9 goes 3 remainder 0.
We then multiply the $\bar{2}$ by 3 to get $\bar{6}$ and this is to be subtracted from 7.
But subtracting a negative number means adding it so we get $7-\bar{6}=13$ for the remainder.

This is much easier and it means that:

> **whenever we use a bar number on the flag we add the product at each step instead of subtracting it.**

Practice F Divide the following, giving answer and remainder:

a 373 ÷ 58 **b** 357 ÷ 48 **c** 300 ÷ 59 **d** 321 ÷ 47

e 505 ÷ 78 **f** 543 ÷ 68

Find to 2 decimal places: **g** 777 ÷ 47 **h** 83.222 ÷ 58 **i** 13 ÷ 79

---

a	**6r25**	b	**7r21**	c	**5r5**	d	**6r39**
e	**6r37**	f	**7r67**				
g	**16.53**	h	**1.43**	i	**0.16**		

Two other variations may be noted.

9 **277 ÷ 38**.

We write 38 as $4\bar{2}$.

	$\bar{2}$	2 7	7
4			$_3$
		6	49

We get 6 remainder 49.
Since we are trying to find how many 38's there are in 277 we cannot allow a remainder greater than 38. There is clearly another 38 in the remainder so the answer, 6, must be increased to 7, and the remainder reduced to 11.

So the answer is **7 remainder 11**.

10 Find **545.45 ÷ 29** to 2 decimal places.

	$\bar{1}$	5 4 5 .4 5 0
3		$_2$ $_1$ $_2$ $_1$ $_1$
		1 8 .7 $_1$0 8

In the third cycle we find we get 31 to divide by 3, and this gives a 2-figure answer.
We simply put the 2-figure answer down (10 remainder 1) and carry on.
The 1 in this 10 is then carried back to the 7 to give 18.808.
So the answer is **18.81** to 2 d.p.

❋ It is possible to avoid this 10 in the answer by having a negative remainder in the previous cycle. You may like to do the sum this way. You should get 18.808 just the same.

Practice G Divide the following, giving answer and remainder:

a 234 ÷ 39 **b** 345 ÷ 49 **c** 555 ÷ 67 **d** 24454 ÷ 37

e 999 ÷ 48 **f** 917499 ÷ 67 **g** 32243 ÷ 48 **h** 24464 ÷ 37

i 81201 ÷ 27 **j** 135791 ÷ 28

a	**6r0**	b	**7r2**	c	**8r19**	d	**660r34**
e	**20r39**	f	**13694r1**	g	**671r35**	h	**661r7**
i	**3007r12**	j	**4849r19**				

Another choice you have is in restructuring the number being divided (the dividend).

11 For example you may want to remove the large digits in the last question of the last exercise by writing 135791 as $136\overline{21}1$.

This is another way of avoiding the 2-figure numbers which can come up in the answer:

$$\begin{array}{c|ccccc|c} \bar{2} & 1 & 3 & 6 & \bar{2} & \bar{1} & 1 \\ 3 & & & {}_1 & {}_0 & {}_2 & {}_0 \\ \hline & 4 & 8 & 4 & 9 & & 19 \end{array}$$

In the following exercise use whatever method you think is best.

Practice H Divide the following, giving answer and remainder:

a 234 ÷ 52 **b** 545 ÷ 83 **c** 343 ÷ 58 **d** 222 ÷ 29

e 141 ÷ 17 **f** 777 ÷ 46 **g** 1344672 ÷ 63 **h** 19792 ÷ 92

i 12585 ÷ 83 **j** 1111 ÷ 19

Divide the following giving answers to 2 decimal places:

k 108 ÷ 31 **l** 4050 ÷ 73 **m** 9876 ÷ 94 **n** 25.52 ÷ 38

o 78 ÷ 49 **p** 6.7 ÷ 88 **q** 19 ÷ 62 **r** 62 ÷ 19

a	**4r26**	b	**6r47**	c	**5r53**	d	**7r19**
e	**8r5**	f	**16r41**	g	**21344r0**	h	**215r12**
i	**151r52**	j	**58r9**				
k	**3.48**	l	**55.48**	m	**105.06**	n	**0.67**
o	**1.59**	p	**0.08**	q	**0.31**	r	**3.26**

11.6 LARGER DIVISORS

Dividing by 3-figure, 4-figure, etc. numbers is an easy extension of the above technique which involves putting all but the first 1 or 2 figures on the flag.

12 **Divide 2250255 by 721**.

We set the sum out with two figures on the flag:

$$\begin{array}{r|ccccccc} 21 & 2 & 2 & 5 & 0 & 2 & 5 & 5 \\ 7 & & & {}^{1} & & & & \\ \hline & & 3 & & & & . & \end{array}$$

Since there are 2 figures on the flag the decimal point goes 2 figures in from the right, as shown.
Then we divide 7 into 22 and put down 3 remainder 1, as shown above.

Next we multiply the answer figure, 3, by the first flag figure, 2, to get 6.
This is deducted from the 15 to give 9 which is divided by 7 to get 1 remainder 2.

$$\begin{array}{r|ccccccc} 21 & 2 & 2 & 5 & 0 & 2 & 5 & 5 \\ 7 & & & {}^{1} & {}^{2} & & & \\ \hline & & 3 & 1 & & & . & \end{array}$$

We now cross-multiply the flag digits, 21, with the last two answer figures, 31, to get 5:

$$\begin{array}{cc} 2 & 1 \\ \multicolumn{2}{c}{\times} \\ 3 & 1 \end{array} \qquad 2 \times 1 + 1 \times 3 = 5$$

This 5 is then deducted from the 20 in the next diagonal to give 15 which is then divided by 7 to give 2 remainder 1.

At every stage from now on we cross-multiply the flag digits with the last two answer digits to get the number to be deducted.

Having got the first 3 answer figures we cross-multiply 21 with 12 to get 5:

$$\begin{array}{cc} 2 & 1 \\ \multicolumn{2}{c}{\times} \\ 1 & 2 \end{array} \qquad 2 \times 2 + 1 \times 1 = 5$$

$$\begin{array}{r|ccccccc} 21 & 2 & 2 & 5 & 0 & 2 & 5 & 5 \\ 7 & & & {}^{1} & {}^{2} & {}^{1} & {}^{0} & {}^{1} \\ \hline & & 3 & 1 & 2 & 1 & .0 & \end{array}$$

Remember the first deduction is the first answer figure times the first flag number. After that we always cross-multiply the flag digits by the last two answer figures.

Practice I Divide the following until you have five figures in the answer:

a 111010 ÷ 725 **b** 17078 ÷ 812 **c** 20006 ÷ 623

d 30405 ÷ 721 **e** 23654 ÷ 713

---

a 153.11 **b** 21.032 **c** 32.112
d 42.170 **e** 33.175

As before we sometimes have to reduce a figure or go into negative numbers. And sometimes we may find it convenient to introduce a bar number into the divisor.

13 For example in dividing **1251413 by 519** we would write 519 as $52\bar{1}$ as this avoids large flag digits and helps to keep the subtractions small.

$2\ \bar{1}$	1	2	5	1	4	1	3	.	0	0	
5			2	1	0	1	0		0	2	
		2	4	1	1	.	2		0	0	

Practice J Divide the following until you have five figures in the answer:

a 253545 ÷ 821 **b** 34567 ÷ 612 **c** 23456 ÷ 621

d 13531 ÷ 629 **e** 7777 ÷ 493 **f** 6789 ÷ 1089 *

---

a 308.82 **b** 56.482 **c** 37.771
d 21.512 **e** 15.775 **f** 6.2341

* Convert 1089 to $11\bar{1}\bar{1}$ and have 11 as a divisor.

"We go on, at last, to the long-promised Vedic process of STRAIGHT (AT SIGHT) DIVISION which is a simple and easy application of the URDHVA-TIRYAK Sutra which is capable of immediate application to all cases and which we have repeatedly been describing as the 'CROWNING GEM of all' for the very simple reason that over and above the universality of its application, it is the most supreme and superlative manifestation of the Vedic ideal of the at-sight mental-one-line method of mathematical computation."

From "Vedic Mathematics", Page 240.

11.7 ALGEBRAIC DIVISION

14 Find $(\mathbf{3x^2 + 10x + 13}) \div (\mathbf{x + 2})$.

$$x + 2\overline{)3x^2 + 10x + 13}$$
$$\underline{\mathbf{3x\ +\ \ 4\ \ rem\ 5}}$$

We divide the first term of the dividend by the first term of the divisor: $3x^2 \div x = \mathbf{3x}$ and put this down.

The coefficient of this 3x i.e. 3 is then multiplied by the +2 in the divisor and the result subtracted in the next column: $3 \times 2 = 6$, and $10 - 6 = \mathbf{4}$, which we put down.

We then multiply this 4 by the +2 in the divisor and subtract the result from the 13: $4 \times 2 = 8$, $13 - 8 = \mathbf{5}$.

15 Find $(\mathbf{x^3 - 5x^2 + 7x - 1}) \div (\mathbf{x - 3})$.

$$x - 3\overline{)x^3 - 5x^2 + 7x\ - 1}$$
$$\underline{\mathbf{x^2 - 2x\ + 1\ rem\ 2}}$$

Similarly each figure brought down into the answer is multiplied by –3 and the result is subtracted in the next column.

16 Find $(\mathbf{x^3 + 6x^2 + 13x + 13}) \div (\mathbf{x^2 + 2x + 3})$.

$$x^2 + 2x + 3\overline{)x^3 + 6x^2 + 13x + 13}$$
$$\underline{\mathbf{x\ + 4\ \ rem\ 2x + 1}}$$

Here we have two coefficients after the first term, +2 and +3.
$x^3 \div x^2 = \mathbf{x}$ which goes into the answer.
Its coefficient, l, then multiplies the +2 and the result is subtracted from the 6 in the divisor: $6 - 2 = \mathbf{4}$.
Now we have the coefficients 1 and 4 in the answer line and we cross-multiply them with the +2 , +3 coefficients in the divisor: $2\times4 + 3\times1 = 11$. This is then subtracted from the 13 in the dividend to give **2x** as the first part of the remainder.

Finally, we multiply the 4 in the answer line with the +3 in the divisor to get 12, which we take from the 13 in the dividend to get **1** as the last part of the remainder.

17 Find $\mathbf{(6x^2 + 17x + 15) \div (2x + 3)}$.

$$2x + 3\overline{)6x^2 + 17x + 15}$$
$$\underline{\mathbf{3x + 4 \quad rem\ 3}}$$

Dividing here by 2x + 3 we proceed as before, but divide each answer figure by 2 before putting it down.
That is: $6x^2 \div 2x = \mathbf{3x}$.
$17 - 3\times3 = 8$, $8 \div 2 = \mathbf{4}$.
$15 - 4\times3 = \mathbf{3}$ (the remainder is not divided).

✏ **Practice K** Divide:

a $x + 5)3x^2 + 20x + 30$

b $x - 3)2x^2 + x - 30$

c $x^2 - 2x - 1)x^3 + 2x^2 + x + 1$

d $x^2 + 2x - 3)2x^3 + 2x^2 + x + 1$

e $2x - 4\ )\ 2x^3 + 2x^2 - 6x + 9$

f $3x + 2)3x^2 + 5x - 1$

---

a	**3x+5**	**R5**	b	**2x+7**	**R –9**	c	**x+4**	**R10x+5**
d	**2x–2**	**R 11x–5**	e	**x^2+3x+3**	**R21**	f	**x+1**	**R–3**

This is very similar to the arithmetic straight division method and other variations (for example, dividing the remainder) can be seen in Chapter 8 of Reference 3.

LESSON 12
EQUATIONS

SUMMARY

12.1 Linear – one-line solutions of linear equations with one or two x terms
12.2 Quadratic Equations – solution using calculus.
12.3 Simultaneous Equations – two special types and the general method.

12.1 LINEAR

12.1a ONE X TERM

The solution of equations comes under the Sutra *Transpose and Apply*. This is equivalent to the common method of "change the side, change the sign". But the Vedic method is to find the answer in one go rather than write down every step of the process.

1 Solve $\mathbf{5x - 4 = 36}$.

Using the Sutra we add 4 to 36 to get 40, then $40 \div 5 = 8$, so $\mathbf{x = 8}$.

2 Solve $\frac{x}{7} + 3 = 5$.

Here we take 3 from 5 to get 2, then multiply 2 by 7, so $\mathbf{x = 14}$.

3 Solve $\frac{2x}{3} = 4$.
Multiply 3 by 4 to get 12, then $12 \div 2 = 6$, so $\mathbf{x = 6}$.

4 Solve $\frac{x-3}{4} = 5$.
Because all the left side is divided by 4 we begin by multiplying 5 by 4, then we add 3 to the result giving $\mathbf{x = 23}$.

Practice A Solve the following equations mentally. Check your answers.

a $3x + 7 = 19$ **b** $2x + 11 = 21$ **c** $3x + 5 = 29$ **d** $7x + 10 = 31$

e $4x - 5 = 7$ **f** $3x - 8 = 10$ **g** $5x - 21 = 4$ **h** $2x - 5 = 6$

i $\frac{x}{3} + 4 = 6$ **j** $\frac{x}{4} + 7 = 9$ **k** $\frac{x}{2} - 8 = 2$ **l** $\frac{x}{3} - 1 = 6$

m $\frac{2x}{3} = 8$ **n** $\frac{3x}{4} = 15$ **o** $\frac{5x}{3} = 15$ **p** $\frac{2x}{5} = 20$

q $\frac{x+4}{7} = 5$ **r** $\frac{x-21}{10} = 1$ **s** $2x + 1 = 3.8$ **t** $3x + 2 = 6.11$

a	**4**	**b**	**5**	**c**	**8**	**d**	**3**
e	**3**	**f**	**6**	**g**	**5**	**h**	**5.5**
i	**6**	**j**	**8**	**k**	**20**	**l**	**21**
m	**12**	**n**	**20**	**o**	**9**	**p**	**50**
q	**31**	**r**	**31**	**s**	**1.4**	**t**	**1.37**

5 Solve $\mathbf{\frac{3x}{5} + 4 = 10}$.

First $10 - 4 = 6$, then $6 \times 5 = 30$, then $30 \div 3 = 10$ so $\mathbf{x = 10}$.

6 Solve $\mathbf{\frac{3x+2}{4} = 8}$.

First $8 \times 4 = 32$, then $32 - 2 = 30$, then $30 \div 3 = 10$ so $\mathbf{x = 10}$.

7 Solve $\mathbf{2(3x + 4) = 38}$.

First $38 \div 2 = 19$, then $19 - 4 = 15$, then $15 \div 3 = 5$ so $\mathbf{x = 5}$.

Alternatively, here, we can multiply the bracket out first:
$6x + 8 = 38$ and then $38 - 8 = 30$ and $30 \div 6 = \mathbf{5}$.

✎ **Practice B** Solve the following:

a $\frac{2x}{3} + 4 = 8$ **b** $\frac{3x}{5} - 4 = 5$ **c** $\frac{7x}{2} - 10 = 11$ **d** $\frac{3x}{8} + 17 = 20$

e $\frac{2x+1}{3} = 4$ **f** $\frac{2x-3}{5} = 3$ **g** $\frac{5x+2}{3} = 9$ **h** $\frac{6x-1}{7} = 5$

i $3(5x - 2) = 54$ **j** $8(x + 3) = 64$ **k** $3(7x - 3) = 33$ **l** $2(4x + 3) = 102$

a	**6**	**b**	**15**	**c**	**6**	**d**	**8**
e	**5.5**	**f**	**9**	**g**	**5**	**h**	**6**
i	**4**	**j**	**5**	**k**	**2**	**l**	**12**

12.1b TWO X TERMS

8 Solve $\mathbf{5x + 3 = 3x + 17}$.

These equations can also be solved mentally. We can see how many x's there will be on the left and what the number on the right will be when we have transposed. Then we just divide the number on the right by the coefficient on the left.

So, in the example above we see there will be 2x on the left when the 3x is taken over, and that there will be 14 on the right when the 3 is taken over.
Then we just divide 14 by 2 to get $\mathbf{x = 7}$.

9 Solve $\mathbf{7 - 2x = x - 5}$.

Here seeing the –2x it is best to collect the x terms on the right.
Then the –2x will become +2x on the right, and the –5 will be +5 on the left.
This gives 12 = 3x (mentally). So $\mathbf{x = 4}$.

10 Solve $\mathbf{2(3x + 4) = 2x + 20}$.

Mentally we see there will be 4x on the left and 12 on the right. So $\mathbf{x = 3}$.

✐ **Practice C** Solve the following mentally:

a $7x - 5 = 4x + 10$ **b** $5 + 4x = 13 + 2x$ **c** $7x + 3 = 15 + x$ **d** $5x - 21 = x - 1$

e $6x + 1 = 4x - 3$ **f** $5x - 1 = 3x + 9$ **g** $8x - 9 = 5x + 12$ **h** $8x - 2 = 6x - 23$

i $10x + 1 = 25 - 2x$ **j** $2(3x + 1) = x + 27$ **k** $3(x - 3) = 2x + 8$ **l** $14 - 3x = x + 10$

m $17 - 5x = 8 - 2x$ **n** $9x - 3 = -x + 57$ **o** $4x - 7 = x + 9$

---

a 5	**b** 4	**c** 2	**d** 5
e –2	**f** 5	**g** 7	**h** –10.5
i 2	**j** 5	**k** 17	**l** 1
m 3	**n** 6	**o** $\frac{16}{3}$	

"The underlying principle behind all of them is Paravartya Yojayet which means: 'Transpose and adjust'. The applications, however, are numerous and splendidly useful."
From "Vedic Mathematics", Page 103.

"In the Vedic mathematics Sutras, CALCULUS comes in at a very early stage."
From "Vedic Mathematics", Page 157.

12.2 QUADRATIC EQUATIONS

There is a simple relationship between the differential and the discriminant in a quadratic equation:

> The differential is equal to the square root of the discriminant.

The **discriminant** of a quadratic expression is defined as follows:

> The discriminant of a quadratic expression is
> **the square of the coefficient of the middle term minus the product of twice the coefficient of first (x^2) term and twice the last term.**

That is, in the quadratic expression $ax^2 + bx + c$, the first differential is $2ax + b$ and the discriminant is $b^2 - 4ac$.

So that $\mathbf{2ax + b = \pm\sqrt{b^2 - 4ac}}$.

This is equivalent to the usual form of the formula for solving quadratic equations, $x = \dfrac{-b \pm \sqrt{b^2 - 4ac}}{2a}$, but is much simpler.

11 Solve $\mathbf{3x^2 + 8x - 3 = 0}$.

The differential is the square root of the discriminant gives $\mathbf{6x + 8} = \pm\sqrt{100}$.
So $\mathbf{6x + 8 = \pm 10}$.
$\mathbf{6x + 8 = 10}$ gives $\mathbf{x = \frac{1}{3}}$. And $\mathbf{6x + 8 = -10}$ gives $\mathbf{x = -3}$.

12 Solve $\mathbf{x^2 - 3x - 1 = 0}$ giving the exact answer.

The differential is the square root of the discriminant gives $2x - 3 = \pm\sqrt{13}$.

Adding 3 on the right and then halving gives an exact answer of $\mathbf{x = \frac{1}{2}(\pm\sqrt{13} + 3)}$.

Practice D Solve:

a $2x^2 + x - 6 = 0$ **b** $x^2 - 2x - 4 = 0$ **c** $x^2 + 5x - 2 = 0$ **d** $2x^2 + 8x + 1 = 0$

---

a $x = \frac{3}{2}, -2$ **b** $x = \frac{1}{2}\left(\pm\sqrt{20} + 2\right)$ **c** $x = \frac{1}{2}\left(\pm\sqrt{33} - 5\right)$ **d** $x = \frac{1}{4}\left(\pm\sqrt{56} - 8\right)$

12.3 SIMULTANEOUS EQUATIONS

12.3a *BY ADDITION AND BY SUBTRACTION*

The formulae *By Addition and By Subtraction* and *By Alternate Elimination and Retention* can be used to solve simultaneous equations.

13 Solve **x + y = 6,**
x − y = 2.

These equations are actually easy to solve because adding them will eliminate y and subtracting them will eliminate x:

$$\begin{array}{l} x + y = 6 \\ \underline{x - y = 2} \; + \\ \underline{2x \quad = 8} \quad \text{so } x = 4 \end{array} \qquad \begin{array}{l} x + y = 6 \\ \underline{x - y = 2} \; - \\ \underline{\quad 2y = 4} \quad \text{so } y = 2. \end{array}$$

So the answer is **x = 4 and y = 2**.

We can also bring in the *Proportionately* formula in order to make further use of the above method.

14 Solve **3x + 2y = 120,**
x + y = 50.

Multiplying the second equation by 2 we get **2x + 2y = 100**.

$$\begin{array}{l} 3x + 2y = 120 \\ \underline{2x + 2y = 100} \; - \\ \underline{\; x \qquad = 20} \end{array}$$

Substituting x=20 into the second equation above gives **y = 30**.

12.3b A SPECIAL TYPE

15 Solve **19x − 7y = 119,**
7x − 19y = 11.

Notice the patterns in the coefficients of x and y here: ax–by=c, bx–ay=d.
When we see such a pattern as this then two applications of *By Addition and By Subtraction* will solve the equations.

Adding the equations gives: $26x - 26y = 130$,
which on division by 26 gives $\mathbf{x - y = 5}$.

Subtracting the equations gives: $12x + 12y = 108$,
which on division by 12 gives $\mathbf{x + y = 9}$.

We now apply addition and subtraction to $x + y = 9$,
$x - y = 5$ to get $\mathbf{x = 7, y = 2}$.

Practice E Solve:

a $4x - 3y = 18$
$3x - 4y = 10$

b $7x - 4y = 48$
$4x - 7y = 18$

c $6x - 5y = 27$
$5x - 6y = 17$

d $9x - 4y = 88$
$4x - 9y = 3$

---

a x=6, y=2 **b** x=8, y=2 **c** x=7, y=3 **d** x=12, y=5

12.3c GENERAL METHOD

Bharati Krsna gives the following general formula for the solution of equations of the form:

$$ax + by = p,$$
$$cx + dy = q.$$

This is $x = \frac{bq - pd}{bc - ad}$, $y = \frac{pc - aq}{bc - ad}$.

16 Solve $\mathbf{2x + 3y = 13}$,
$\mathbf{5x + 2y = 16}$.

Therefore $x = \frac{3 \times 16 - 13 \times 2}{3 \times 5 - 2 \times 2} = 2$, $y = \frac{13 \times 5 - 2 \times 16}{3 \times 5 - 2 \times 2} = 3$.

This is similar to the current method which uses determinants.

Practice F Solve:

a $3x + 5y = 19$
$2x + 3y = 12$

b $2x + 3y = 14$
$5x + 7y = 33$

c $2x + 5y = 19$
$3x + 2y = 12$

d $3x + 2y = 5$
$7x + 3y = 10$

---

a 3, 2 **b** 1, 4 **c** 2, 3 **d** 1, 1

12.3d ANOTHER SPECIAL TYPE

17 Solve **$3x + 2y = 6$,**
$9x + 5y = 18$.

This comes under the Vedic Sutra *If One is in Ratio the Other One is Zero.*
We notice that the ratio of the x coefficients is the same as the ratio of the coefficients on the right-hand side: **3:9 = 6:18**.
This tells us that since x is in ratio the other one, y, is zero: **y = 0**.
If y=0 we can easily find x by putting y=0 in the first (or the second) equation:
$3x + 0 = 6$.

Therefore x=2. So **x = 2, y = 0**.

So look out for where the ratio of the x or y coefficients is the same as the ratio on the right: the other one is then zero.

Practice G Solve:

a $3x + 2y = 6$ $2x + 3y = 9$	**b** $3x + 8y = 6$ $5x + 3y = 10$	**c** $4x - y = 20$ $x + 5y = 5$	**d** $13x + 17y = 2$ $9x + 51y = 6$

---

a **x=0, y=3** b **x=2, y=0** c **x=5, y=0** d **x=0, y=$\frac{2}{17}$**

Variations of this type are shown in Reference 3.

There are many other special types of equation in the Vedic system and spotting them can save us a lot of time and effort.

"Well, I may say at this point that whatever kind of question there may be, in whatever chapter it may be, simple equations, quadratic equations, cubic, bi-quadratic, and various other subjects of algebra, geometry, plane trigonometry, spherical trigonometry, calculus, and so on, the question must come under one of the sixteen rules. All that we have to do is, let me repeat, to recognize and identify the particular type and apply the Sutra relative to that particular type.
From "Vedic Metaphysics", Page 180.

LESSON 13
APPLICATIONS OF TRIPLES

SUMMARY

13.1 Triple Subtraction
13.2 Triple Geometry – working with triples rather than angles.
13.3 Angle Between Two Lines – getting a triple representing this.
13.4 Half Angle – obtaining a triple with half the angle of a given triple.
13.5 Coordinate Geometry – gradients, solving circle problems and finding the length of a perpendicular from a point onto a line.

13.1 TRIPLE SUBTRACTION

1 If **A) 4 3 5** and **B) 15 8 17** find **a triple for A–B**.

The diagram below shows how the angles are subtracted: the base of 15, 8, 17 is placed as usual on the hypotenuse of the 4, 3, 5 triangle. But its hypotenuse is below the base so that the angle B is subtracted from A.

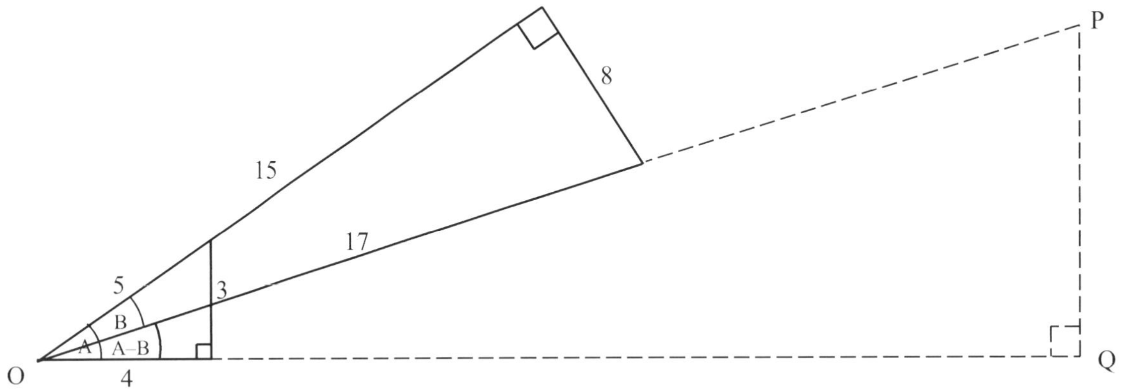

The vertical and crosswise pattern is exactly as before, the difference being that where before we subtracted we now add, and where before we added we now subtract:

A	4	3	5
B	15	8	17
A–B	$(4 \times 15 + 3 \times 8)$,	$(3 \times 15 - 4 \times 8)$,	(5×17)
=	84	13	85

Alternatively we may think of triple subtraction as adding a triple with a negative height:

A	4	3	5	
B	15	–8	17	+
A–B	84	13	85	as before.

2

3	4	5	
4	3	5	–
24	7	25	

3

A	4	3	5	
B	3	4	5	–
A–B	24	–7	25	

Here we get a triple with a negative height. This is because B , the angle being subtracted, is larger than the angle A. See the diagram below.

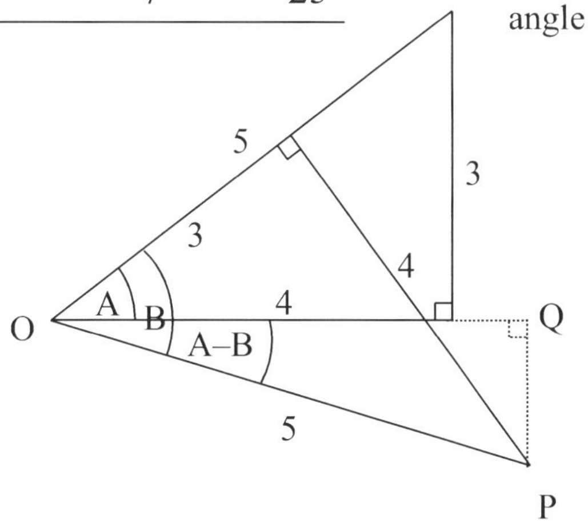

4 If **A) –3 4 5** and **B) 12 5 13** find a triple for **2A–B**

A	–3	4	5	
2A	–7	–24	25	
B	12	5	13	–
2A–B	–204	–253	325	

Here 2A and 2A–B are both in the third quadrant.

✎ **Practice A** Subtract the following triples:

a	7 24 25	**b**	4 3 5	**c**	12 5 13	**d**	12 5 13	**e**	3 –4 5
	3 4 5 –		12 5 13 –		24 7 25 –		3 4 5 –		3 4 5 –

Given A)12,5,13 B)5, 12, 13 C)3,–4,5 find triples for:

f A–B **g** B–C **h** 2C–B **i** A–B–C **j** A–A

a	**117, 44, 125**	**b**	**63, 16, 65**	**c**	**323, 36, 325**	**d**	**56, –33, 65**	**e**	**–7,–24, 25**
f	**120,–119, 169**	**g**	**–33, 56, 65**	**h**	**–323,–36, 325**	**i**	**836, 123, 845**	**j**	**169, 0, 169 = 1,0,1**

5 **Given A) 4 3 5 find a triple for 180°– A**.

180°	–1	0	1	
A	4	3	5	–
180°–A	–4	3	5	

That is we simply change the sign of the first element of a triple to obtain the **supplementary triple**.

6 **Given A) 3, 4, 5 find a triple for –A**.

This means that A is measured clockwise rather than anti-clockwise.

Since 0 – A = –A we can get the required triple by subtracting the triple from A from the triple for 0°:

0°	1	0	1	
A	3	4	5	–
–A	3	–4	5	

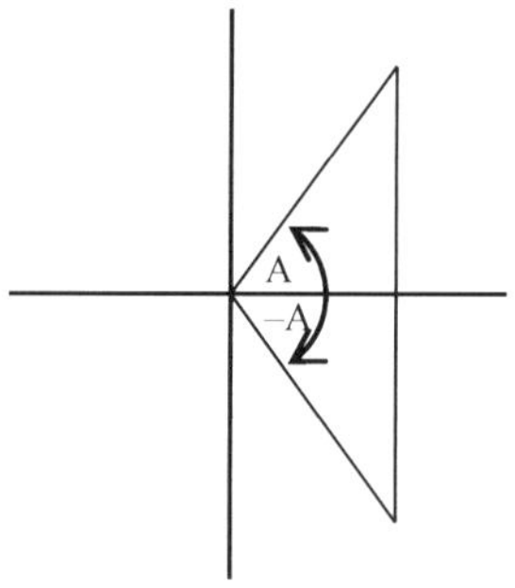

In general we change the sign of the middle element of a triple A to obtain the triple for –A. Geometrically the triangle is simply reflected in the horizontal axis.

7 **Rotate P(4,–5) clockwise through A) 4,3,5 about O**.

Rotations were introduced in Lesson 9 where we added triples for anti-clockwise rotations. For clockwise rotations we subtract the triples.

4	–5	-	
4	3	5	–
1	–32	-	

$\therefore$ P'(0.2, –6.4).

✎ **Practice B** Find the following triples:

a Given A) 12, 5, 13 find A–90°.
b Given A) 15, 8, 17 find A–180° .
c Given A) 5, 12, 13 find triples for:
i) –A,
ii) 45°– A,
iii) 270°– 2A.

d Find a triple for $-15°$.

e Rotate (0,–3) clockwise about the origin by the triple 3, 4, 5.

f Through what angle (triple) must the point (10,3) be rotated about the origin so that it is in the direction OP where P is the point (5,4)?

a 5,–12, 13	**b –15, –8, 17**	
c i) 5, –12, 13	**ii) 17, –7, 13 $\sqrt{2}$**	**iii) –120, 119, 169**
d $\sqrt{3}+1, 1-\sqrt{3}, 2\sqrt{2}$	**e (–2.4, –1.8)**	**f 62, 25, -**

13.2 TRIPLE GEOMETRY

We can solve triangles etc. by working with triples rather than angles.

8 Two angles of a triangle are given by the triples **4, 3, 5** and **5, 12, 13**.
Find the triple for the third angle.

Given two angles of a triangle we would normally add them and take the result from 180°.
Here we do the same thing with the triples:

$$\begin{array}{rrr} 4 & 3 & 5 \\ 5 & 12 & 13 + \\ \hline -16 & 63 & 65 \\ \hline \end{array}$$

We can subtract this result from the triple for 180° or use the short cut suggested in Example 5 to get **16, 63, 65** for the answer.

Later we will see how to get this answer more easily.

9 **Find a triple for angle CAD**:

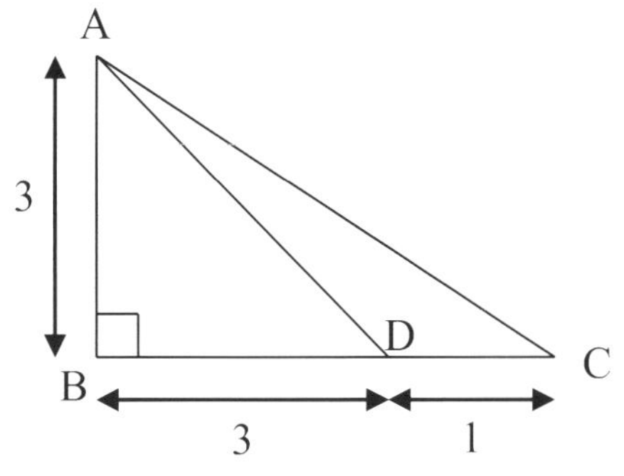

Since the required angle is the difference of the angles CAB and DAB we can write down triples for CAB and DAB and subtract them to get a triple for angle CAD.

CAB:	3	4	5
DAB:	1	1	$\sqrt{2}$ –
CAD:	7	1	$5\sqrt{2}$

Practice C

a Given that in triangle ABC, AB=AC and that a triple for B is B)12, 5, 13 show that the triple for angle BAC is –119, 120, 169.

b Find a triple for angle CAD:

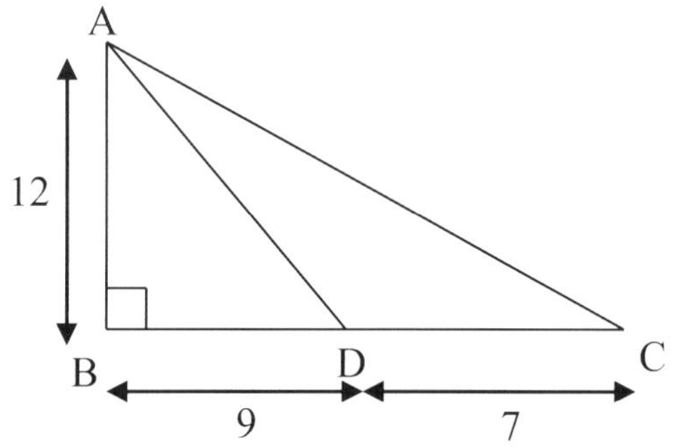

c Find a triple for angle x:

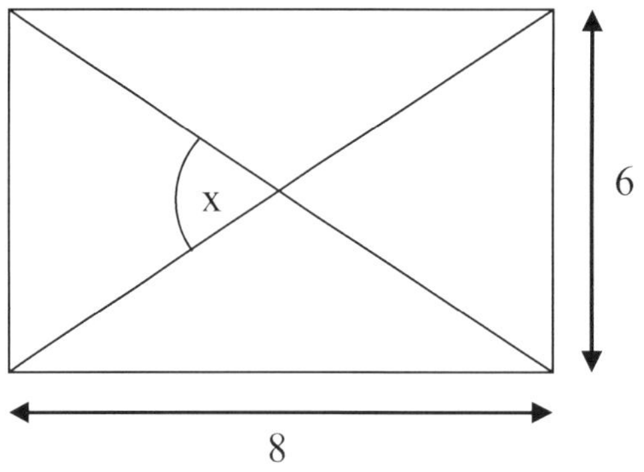

b **24, 7, 25** **c** **7, 24, 25**

13.3 ANGLE BETWEEN TWO LINES

10 **Find the angle between the lines 3y = 4x – 12 and y = x + 10.**

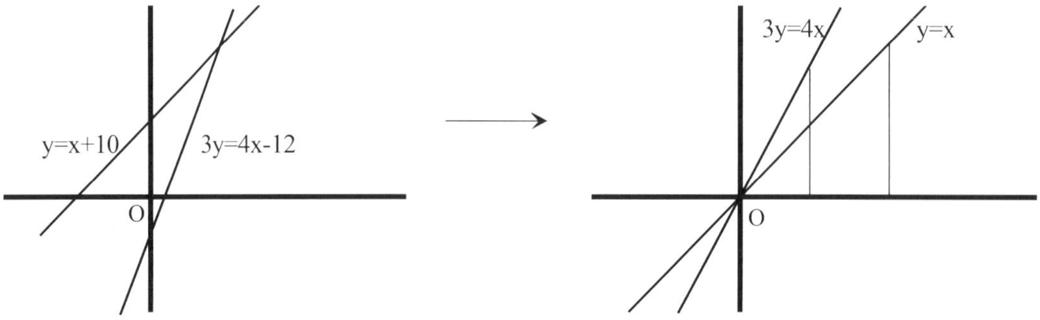

We are interested here only in the slope of the lines and so can disregard the absolute terms –12 and 10.

Then subtracting the line triples gives a triple for the angle between the two lines:

3	4	5	
1	1	$\sqrt{2}$	–
7	1	$5\sqrt{2}$	which represents the required angle.

11 **Find the angle between y = 2x + 9 and 2y = –x – 4.**

1	2	$\sqrt{5}$	
2	–1	$\sqrt{5}$	–
0	5	5	= 0, 1, 1

The lines are therefore perpendicular.

12 **Find the angle between y = 2x + 1 and 3y = x.**

1	2	$\sqrt{5}$	
3	1	$\sqrt{10}$	–
5	5	$\sqrt{50}$	= 1, 1, $\sqrt{2}$

The angle is therefore 45°.

The angles between pairs of lines in this section have been left in triple form. Finding the angle in a given triple and a triple for a given angle is not dealt with here (see "Triples").

Practice D Find the angle between the lines:

a y = 3x, y = 2x

b 2y = 3x, 3y = 4x

c 2y = 3x + 4, y = x + 3

d 2y = 3x, 5y = x

e 2y = 3x, 3y = 2x

f 3y = 2x, 2y + 3x = 0

---

a **7, 1, $\sqrt{50}$** b **18, 1, $5\sqrt{13}$** c **5, 1, $\sqrt{26}$**
d **45°** e **12, 5, 13** f **90°**

13.4 HALF ANGLE

Here we see how to find triples containing half the angle in a given triple, and we consider only the case where 0°<A<180°.

13 **Given A) 7 24 25 find a triple for ½A.**

½A) 7+25, 24, - = 32, 24, - = **4, 3, 5**

That is, we add the first and last elements of the triple to get the first element of the half-angle triple and keep the middle element as the middle element of the half-angle triple.
The third element can be calculated from the first two.

Proof:

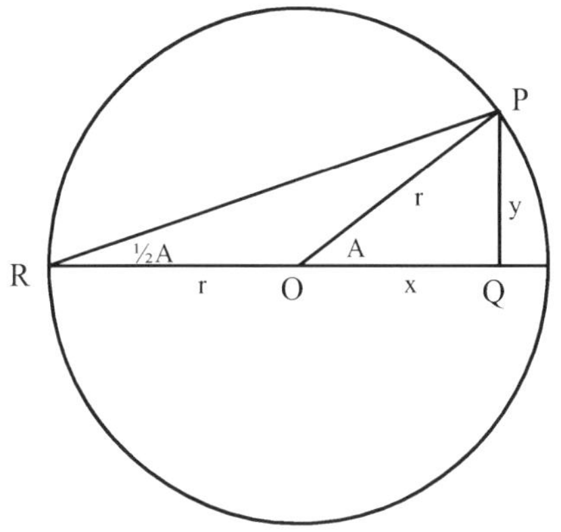

Given a triple A) x, y, r draw a circle centre O, radius r, as shown.
Produce QO to R and join RP.
Then the angle at R is ½A (a well-known circle theorem: the angle subtended at the centre is twice that at the circumference) so that from ΔRPQ the triple for ½A is ½A) x+r, y, -.

Similarly if A is obtuse, so that x is negative:

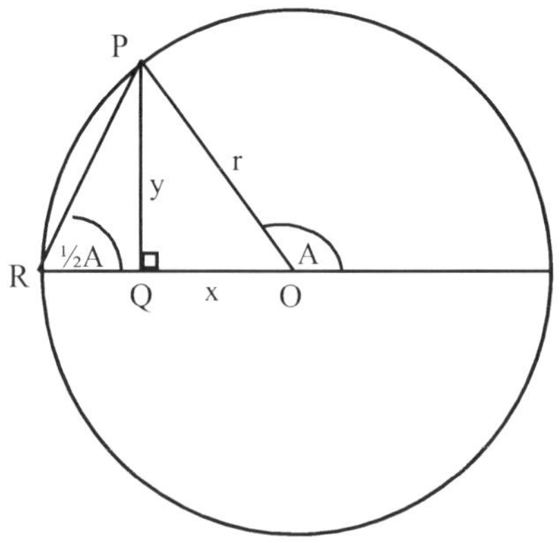

We still have ½A) x+r, y, -

14 If **A) 4, 3, 5** then ½A) 4+5, 3, - = **3, 1, $\sqrt{10}$**.

15 If **A) –3, 4, 5** then ½A) 2, 4, - = **1, 2, $\sqrt{5}$**.

16 In the diagram the angle A is given by the 3, 4, 5 triple.

Find a triple for the obtuse angle at C.

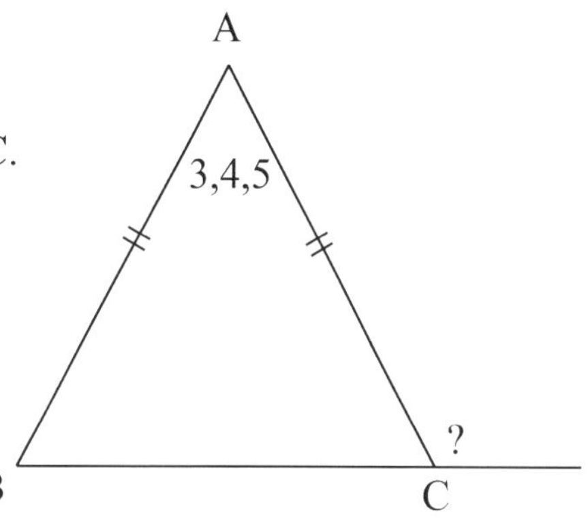

We could subtract 3,4,5 from the 180° triple, halve the result and then subtract from 180°.

	180°	–1	0	1	
	A	3	4	5	–
subtract 3,4,5 from the 180° triple	180°–A	–3	4	5	
halve the result	½(180°–A)	1	2	$\sqrt{5}$	= B = C
subtract from 180°	180°–C	–1	2	$\sqrt{5}$	Answer

Alternatively we can halve the isosceles triangle as shown, halve the 3,4,5 triple to get 2, 1, -, for the angle CAD,

find the complementary triple 1, 2, -, for the angle ACD

and then the supplementary triple **–1, 2, -**.

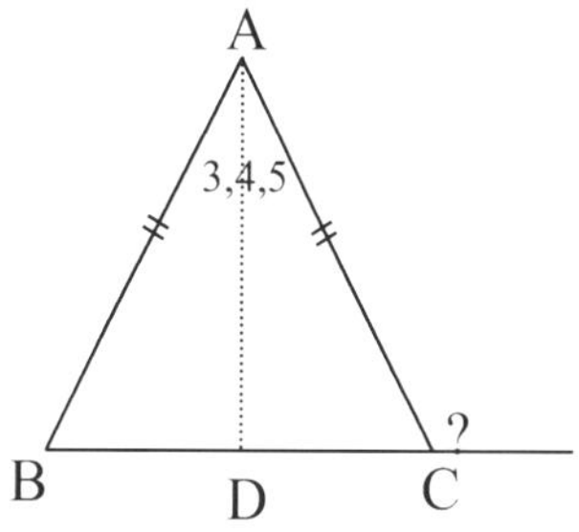

Practice E Find the half angle triple for:

a 5,12,13 **b** 15,8,17 **c** 8,15,17 **d** 119,120,169 **e** –4,3,5

f –119,120,169 **g** –7,24,25

h Given A) 3,4,5 find triples for: i) ½A
ii) ½A+A

i Find triples for: i) 15° ii) 22½° iii)285°.

j Find a triple for angle x:

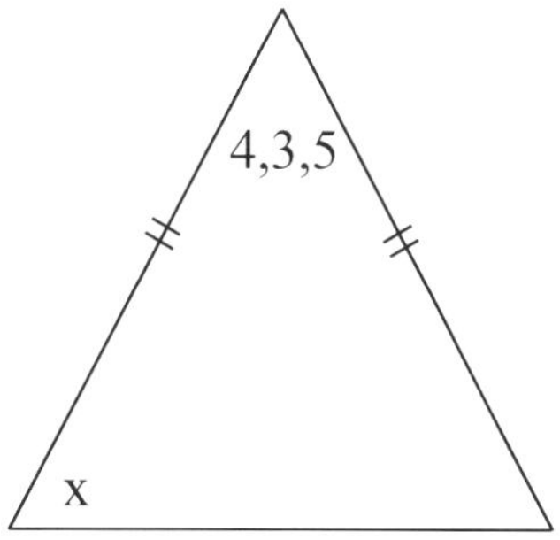

k In the diagram, CB is a tangent to the circle, angle AOB is given by the triple –3,4,5.

Find triples for angles OCB and ABD.

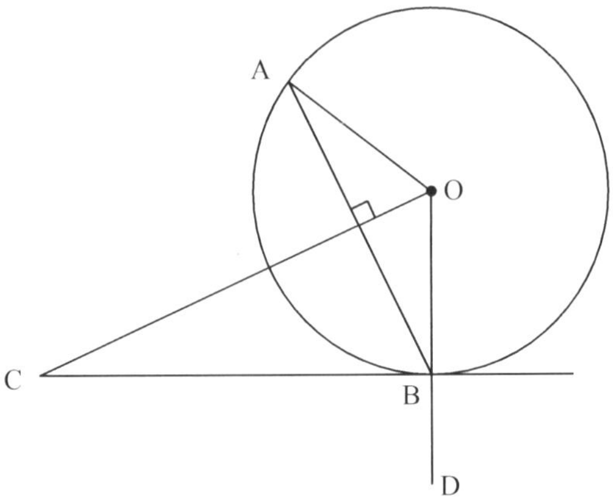

---

a 3, 2, $\sqrt{13}$	**b** 4, 1, $\sqrt{17}$	**c** 5, 3, $\sqrt{34}$	**d** 12, 5, 13	**e** 1, 3, $\sqrt{10}$
f 5, 12, 13	**g** –3, 4, 5			
h i) 2, 1, $\sqrt{5}$	**ii)** 2, 11, $5\sqrt{5}$	**i i)** $\sqrt{3}+2$, 1, - or $\sqrt{3}+1$, $\sqrt{3}-1$, $2\sqrt{2}$		**ii)** $\sqrt{2}+1$, 1, -
iii) 1, $-\sqrt{3}-2$, -	**j** 1, 3, -	**k** 2, 1, -, -2, 1, -		

13.5 COORDINATE GEOMETRY

13.5a GRADIENTS

Triples are extremely useful in many areas of mathematics, including coordinate geometry which deals with forms within a 2-dimensional coordinate system.

This is, in part, because a triple can be used to describe different things, including an angle, a triangle, the position of a point and a gradient.

For example, the line 4y = 3x has a gradient of $\frac{3}{4}$ and a 4,3,5 triangle shows a slope with a gradient of $\frac{3}{4}$ if its base is extended to the right:

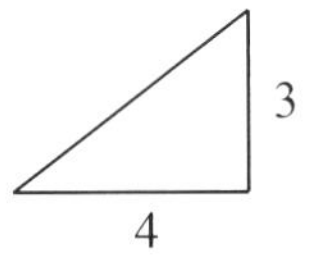

Similarly 12y = 5x will have a gradient of $\frac{5}{12}$ and so its gradient can be represented by the triple 12,5,13.

A triple can describe a point because the coordinates of a point define a right-angled triangle.

So (6,8) is described by the triple 6,8,10:

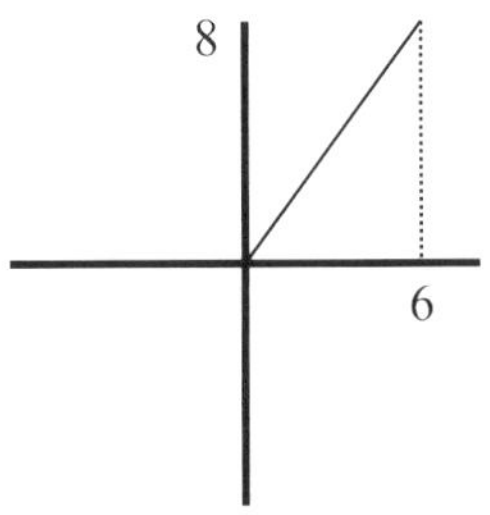

and (–3,2) by $-3, 2, \sqrt{13}$:

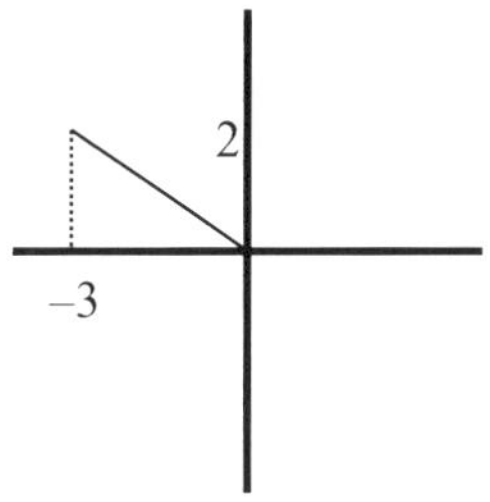

17 **Find the equations of the lines through (5,2) which make an angle of 45° to the line y = 2x + 3.**

There are two possibilities as the 45° could be added to or subtracted from the slope of the line:

line:	1	2	-	
45°	1	1	-	+
	–1	3	-	

$\therefore\ -y = 3x + c$

line:	1	2	-	
45°	1	1	-	–
	3	1	-	

$\therefore\ 3y = x + c$

and since the lines pass through the point (5,2) we get:

–y = 3x – 17 and **3y = x + 1**.

13.5b LENGTH OF PERPENDICULAR

18 **Find the length of the perpendicular from the point P(5,2) onto the line 4y − 3x.**

This is the same as finding the distance of the point (5,2) from the line 4y = 3x.
We require the distance PQ in the diagram below.

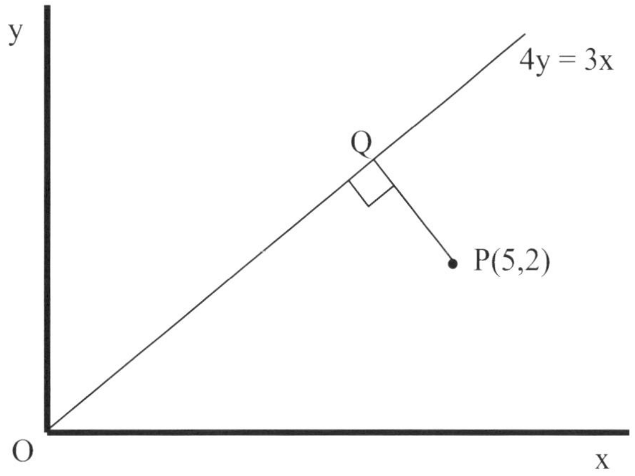

The slope of the line is described by the triple 4,3,5 and the point P can be described by the triple $5, 2, \sqrt{29}$.

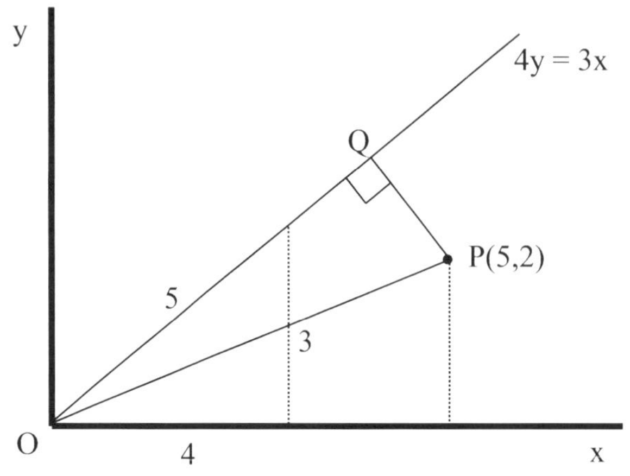

Drawing in the point triple and the gradient triple you can see that by subtracting these triples we get a triple for ΔOPQ:

line triple:	4	3	5	
point triple:	5	2	$\sqrt{29}$	–
	26	7	$5\sqrt{29}$	
divide by 5	$\frac{26}{5}$	$\frac{7}{5}$	$\sqrt{29}$	= ΔOPQ

Since OP = $\sqrt{29}$ we have to divide the third line through by 5 to make the triangle the right size.

$\therefore$ PQ = $\frac{7}{5}$ = **1.4**.

Note:
1. we require only the middle element of the triple for ΔOPQ so we do not need to calculate the others,
2. subtracting the line triple from the point triple would give the same result but with the sign different, therefore we need not trouble ourselves about which triple to subtract from which.

19 **Find the distance of the point (–2,–5) from the line 5y + 12x = 0**.

Think of the equation of the line as 5y = –12x to get the line triple: 5,–12,13.

5	–12	13	
–2	–5	-	–
-	49	-	

$\therefore$ Distance = $\frac{\mathbf{49}}{\mathbf{13}}$.

20 **Find the distance of (6,3) from 3y = 4x + 2.**

Here the line does not pass through the origin.
But we observe that the point (1,2) lies on the line and transpose the origin to there.
Then find the distance of (5,1) from 3y = 4x:

3	4	5	
5	1	-	–
-	17	-	

$\therefore$ Distance = $\frac{17}{5}$ = **3.4**.

No adjustment is required due to the transposition of the origin.

21 **Find the distance of 3x – 2y = 5 from the origin.**

We can transpose the origin to the point (1,–1), which is on the line and then find the distance of (–1,1) from 3x – 2y = 0:

2	3	$\sqrt{13}$	
–1	1	-	–
-	–5	-	

$\therefore$ Distance = $\mathbf{\frac{5}{\sqrt{13}}}$.

Practice F Find the length of the perpendicular from:

a (–3,2) onto 3y = 4x
b (0,4) onto 24y = 7x
c (9,10) onto 3y = 4x + 3
d (3,5) onto 6y + 8x = 3
e (1,1) onto 12y = 5x – 13
f (4,–5) onto y = 3x – 4
g (6,2½) onto 4y = 3x
h (4,5) onto y = 2x + 3
i Find the shortest distance of the line y = 3x from the circle $(x-4)^2 + (y-3)^2 = 4$.

---

a **3.6** b **3.84** c **1.8**
d **5.1** e $\frac{20}{13}$ f $\frac{13}{\sqrt{10}}$
g **1.6** h $\frac{6}{\sqrt{5}}$
i $\frac{9}{\sqrt{10}} - 2$

13.5c CIRCLE PROBLEMS

22 **Find the coordinates of the points of contact of the tangents to the circle $(x-3)^2 + (y-4)^2 = 9$ which pass through the origin.**

The centre of the circle is at (3,4) and the radius is 3.

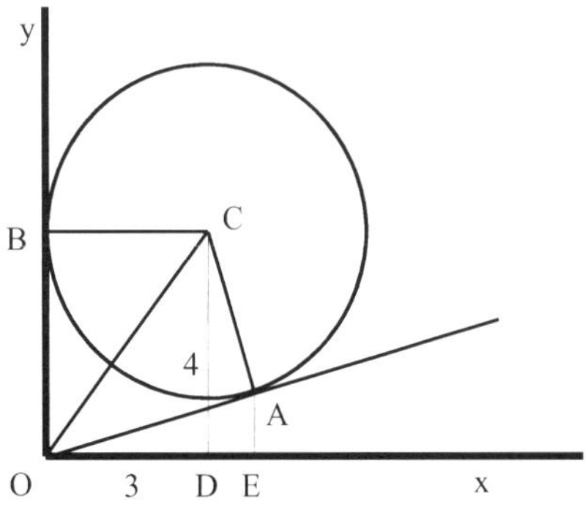

So OC = 5

and OA = OB = 4.

For A we obtain ΔOEA by subtracting ΔOAC from ΔODC:

	3	4	5	
	4	3	5	–
angle AOE	24	7	25	

$\therefore$ as OA = 4 the triple for angle AOE must be multiplied by $\frac{4}{25}$ i.e. $\frac{96}{25}, \frac{28}{25}, 4$.

$\therefore$ A$\left(\frac{96}{25}, \frac{28}{25}\right)$ And B is clearly at **(0,4)**.

23 **Find the equation of the circle, centre (3,4), to which the line y = 2x is tangential.**

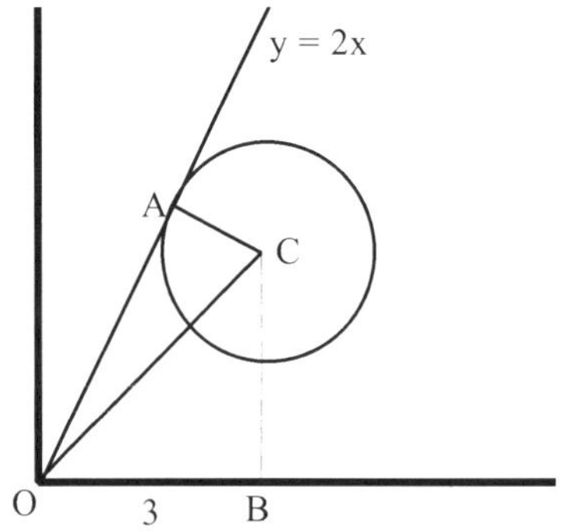

We need the radius of the circle, AC.
Since the sum of the two triangles shown has a gradient of 2 we can find ΔOAC by triple subtraction:

	1	2	$\sqrt{5}$	
	3	4	5	−
angle AOC	11	2	$5\sqrt{5}$	$= \frac{11}{\sqrt{5}}, \frac{2}{\sqrt{5}}, 5$, since OC = 5.

$\therefore$ AC = $\frac{2}{\sqrt{5}}$ and the equation of the circle is $\mathbf{(x-3)^2 + (y-4)^2} = \frac{4}{5}$.

24 **Find the equation of a circle of radius 4 which touches the x-axis and the line 4y = 3x.**

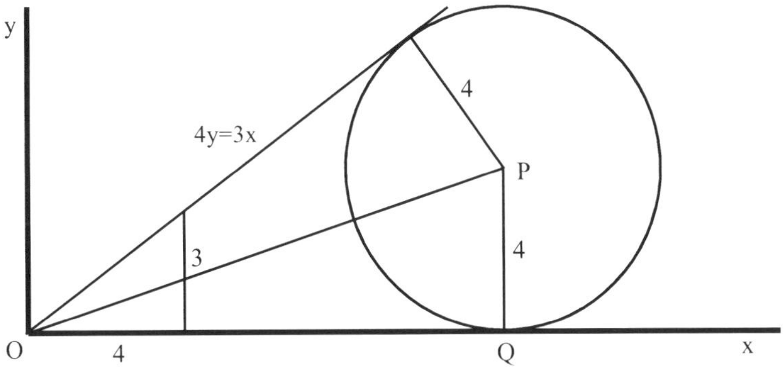

The line triple is 4,3,5 and so the half angle triple is 3,1, -.
$\therefore$ as OP bisects the angle formed by the two tangents ΔOPQ is represented by the triple 3, 1, -.
But since PQ=4 the coordinates of P are (12,4).
The equation of the circle is then $\mathbf{(x - 12)^2 + (y - 4)^2 = 4^2}$.

25 **Find the coordinates of the centre of a circle, radius 6, which touches the lines 3y = 4x and 12y = 5x.**

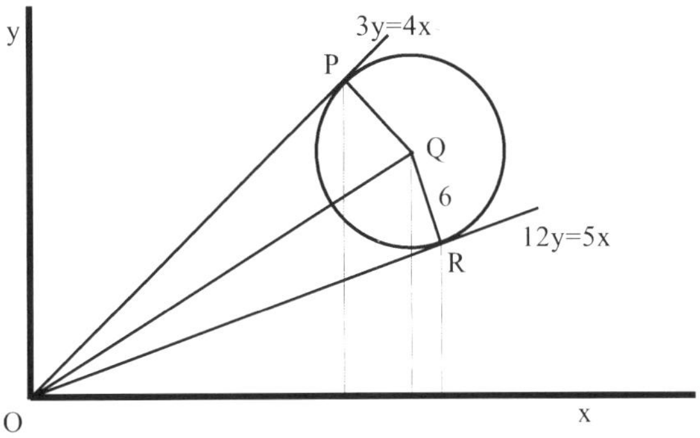

By subtracting the triples for the two given lines we get a triple for angle POR.
The half angle triple then gives us a triple for angle QOR.
We can then add this to the triple for 12y = 5x to get a triple for angle QOx.

So:	3	4	5	
	12	5	13	−
	56	33	65	triple for $P\hat{O}R$

Halving this we get: 11,3,- as a triple for $Q\hat{O}R$.
And since QR = 6, OR = 22.

$$\therefore \quad \begin{array}{ccc} 22 & 6 & OQ \\ 15 & 5 & 13 \\ \hline 234 & 182 & 13OQ \end{array} + \quad = 18,\ 14,\ OQ.$$

The coordinates of Q are therefore **(18,14)**.

Practice G

a Find the equations of the lines through the point (3,1) which make an angle of 45° to the line $y = 3x + 1$

b Find the equation of the circle, centre (2,5), to which the line $y = 3x$ is tangential.

c Find the equation of the upper tangent from the origin to the circle:
$(x - 24)^2 + (y - 7)^2 = 15^2$.

---

a **$y=-2x+7$, $2y=x-1$** b **$(x-2)^2 + (y-5)^2 = 0.1$** c **$3y = 4x$**

There are many other applications of triples: see Reference 4. In Manual 3 we show applications in trigonometry (proving trig. identities, solving trig. equations etc.), coordinate geometry, complex numbers, simple harmonic motion, projectile motion etc.

LESSON 14
SQUARE ROOTS

SUMMARY

14.1 Squaring – revision.
14.2 Square Root of a Perfect square – where the square root is a 2 or 3-figure number.
14.3 General Square Roots
14.4 Changing the Divisor – choosing a small or large divisor.

14.1 SQUARING

The general method of finding the square root of a number is just the reverse of the squaring process so we begin this chapter by revising squaring.

We square a number by combining the Duplexes contained in the number.

1 Find $\mathbf{5431^2}$.

The Duplexes are:
D(5)=**25**, D(54)=**40**, D(543)=**46**, D(5431)=**34**, D(431)=**17**, D(31)=**6**, D(1)=**1**.

Working from left to right we get $2_59_04_69_45_7761$ (see Section 7.1e).

Practice A Square the following numbers, from left to right:

a 23	**b** 34	**c** 54	**d** 61	**e** 421	**f** 124	**g** 423
h 818	**i** 4321	**j** 6032	**k** 5234			

---

a	**529**	b	**1156**	c	**2916**	d	**3721**	e	**177241**	f	**15376**	g	**178929**
h	**669124**	i	**18671041**	j	**36385024**	k	**27394756**						

14.2 SQUARE ROOT OF A PERFECT SQUARE

14.2a PREAMBLE

If we are given a number which we are to find the square root of there are two important facts we can immediately get from the number:

1) the number of figures in the square root before the decimal point,
2) the first figure of the square root.

2 Suppose we want the **square root of 543200**.

We mark off pairs of digits from the right (in fact from the decimal point): 54'32'00. Since there are 3 groups of digits formed there will be 3 figures in the square root before the decimal point.

Since the group on the left is 54 this tells us that the first figure will be **7** because **the first square number below 54 is 49** ($=7^2$).

Using these two results together we can say that since the answer starts with 7 and has 3 figures before the decimal point, $\sqrt{543200} \approx \mathbf{700}$.

3 Find an approximate value for $\sqrt{\mathbf{543.2}}$.

Split the number into pairs starting at the decimal point: 5'43.2.
There are 2 groups (the single digit, 5, counts as a group as 5 = 05) so there will be 2 figures in the answer before the point.

The group at the left is 5 and the first square number below 5 is 4, which is 2^2.
So the first figure of the square root is **2**.

So the square root begins with 2 and has 2 figures before the point.
Therefore $\sqrt{\mathbf{543.2}} \approx \mathbf{20}$.

14.2b TWO-FIGURE ANSWER

4 Find the **square root of 1849**.

Marking off two figures from the right, 18'49, we expect two figures before the decimal point and the first figure of the answer is 4.

We set the sum up like this:

	1	**8**	**4**	**9**
8)		2		
		4		

Since $4^2 = 16$ and 18 is 2 more than this we have a remainder of 2 which we place as shown. **Note the 24 formed diagonally by this 2 and the 4 above it.**
The answer goes on the bottom line.

We also put **twice the first figure**, which is 8, as a divisor at the left as shown.

Next we divide the 24 by the divisor 8.
This gives 3 remainder 0, placed as shown below:

	1	**8**	**4**	**9**
8)		2	0	
		4	**3**	

We now see 09 and we deduct from this the Duplex of the last answer figure:
D(3) = 9 and 09 – 9 = 0. This means the answer is exactly **43**.

5 Find $\sqrt{\mathbf{1369}}$.

Again we expect 2 figures before the point and the first figure will be 3.

	1	**3**	**6**	**9**
6)		4		
		3		

Since $3^2 = 9$ we have a remainder of 4 placed as shown above. Also we again put twice the first figure, 6, as a divisor on the left.
Note the 46 in the diagonal of figures.

Next we divide the divisor, 6, into 46 and put down 7 remainder 4:

	1	**3**	**6**	**9**
6)		4	4	
		3	**7**	

Finally we see 49 in the second diagonal and we take the Duplex of the last answer figure, D(7) = 49, from this to get 0.

So the answer is exactly **37**.

A. We first set up the initial sum including the first figure of the answer, the remainder and twice the first figure placed as a divisor on the left.

B. We then divide the figures shown in the diagonal by the divisor and put down the answer and remainder.

C. Finally check the answer is exact by subtracting the Duplex of the last answer figure from the second diagonal.

✏ **Practice B** Find the square root of the following, using the method shown:

a 3136 **b** 3969 **c** 5184 **d** 3721 **e** 6889 **f** 1296

g 2304 **h** 4624 **i** 1521 **j** 3844 **k** 8649

a	56	b	63	c	72	d	61	e	83	f	36
g	48	h	68	i	39	j	62	k	93		

REVERSING SQUARING

Considering the square root of 1849, the first figure is clearly 4, and there is a remainder of 2, as $4^2 = 16$ and we have 18 in the first two places.

So we could write: $\sqrt{18_249} = 4p$, where 4p is the 2-figure answer, and the remainder, 2, is written as a subscript.

Squaring the 4 (which represents 40) therefore accounts for 1600 of 1849 and so 249 are left (as you can see under the square root above).

Now the duplex of (4p) must account for the 24 in $_249$, or most of it. So twice the product of 4 and p must be 24 or nearly 24. Therefore p must be 3.

This shows why we use twice the first figure as a divisor, since to solve $2\times(4\times p) = 24$ we can divide 24 by 8: we will always be dividing by twice the first digit.

Square roots could be taught this way initially if understanding the steps is important, or of course it can be explained this way afterwards.

"And we were agreeably astonished and intensely gratified to find that exceedingly tough mathematical problems (which the mathematically most advanced present day Western scientific world had spent huge lots of time, energy and money on and which even now it solves with the utmost difficulty and after vast labour and involving large numbers of difficult, tedious and cumbersome "steps" of working) can be easily and readily solved with the help of these ultra-easy Vedic Sutras (or mathematical aphorisms) contained in the Parishishta (the Appendix-portion) of the ATHARVAVEDA in a few simple steps and by methods which can be conscientiously described as mere "mental arithmetic".

From "Vedic Mathematics", Page xv.

14.2c THREE-FIGURE ANSWER

6 Find the **square root of 293764**.

We first mark off pairs of figures from the right: 29'37'64.
This shows us that we expect 3 figures before the decimal point, and that the answer begins with a 5.

We set up the sum as before:

	2	**9**	**3**	**7**	**6**	**4**
10)		4				
		5		.		

We have a remainder of 4 from the 29 and twice the first figure is 10.
Since we know there are 3 figures before the point we can insert the decimal point as shown.

We get the next figure of the answer by dividing 10 into the 43.
This gives 4 remainder 3 which we write down as shown below:

	2	**9**	**3**	**7**	**6**	**4**
10)		4	3			
		5	**4**	.		

Next, before we divide 10 into the 37 in the second diagonal we subtract the duplex of the 4 (in the answer) from it. The duplex of 4 is 16.
37 – 16 = 21 and 21 ÷ 10 = 2 remainder 1, which we write down as shown below:

	2	**9**	**3**	**7**	**6**	**4**
10)		4	3	1		
		5	**4**	**2.**		

Next, before dividing 10 into the 16 in the third diagonal we deduct the duplex of the 42 in the answer. D(42)=16, 16–16=0 and 0÷10=0 rem 0:

	2	**9**	**3**	**7**	**6**	**4**
10)		4	3	1	0	
		5	**4**	**2.**	**0**	

Finally we deduct the duplex of the last answer figure, 2, from the 04 in the fourth diagonal. This leaves 0 and so the answer is exactly **542**.

The method is a continuation of the shorter sums done before.

We set out the initial sum as before and divide into the first diagonal as before.
The next three steps involve deducting the Duplex of **a** the 2nd answer figure,
b the 2nd and 3rd answer figures,
c the 3rd answer figure
from the last diagonal before dividing by the divisor.

Practice C Find the square root of:

a 186624	**b** 264196	**c** 400689	**d** 318096
e 119025	**f** 524176	**g** 59049	**h** 197136
i 519841	**j** 375769	**k** 53361	

---

a	432	b	514	c	633	d	564
e	345	f	724	g	243	h	444
i	721	j	613	k	231		

REVERSING SQUARING

Again, thinking back to how we square a 3-figure number we can reverse the procedure to obtain the square root of, say, 293764.

We know the answer will have three figures and will start with 5. There will therefore be a remainder of 4 (29 – 25 = 4):

$\sqrt{29_43764} = 5pq$, where p and q are the remaining unknown digits.

Now we know that the duplex of 5p = D(5p) = 2×(5×p) must equal 43 ($_4$3 under the square root above) or most of it. Therefore p = 4 and 3 are remaining. In fact we can just divide 43 by 10 (twice the first digit).

So we now have: $\sqrt{29_43_3764} = 54q$.
Next we have our eye on the 37 ($_3$7).
This must be accounted for by the duplex of 54q = D(54q) = 2×(5×q) + 4^2.
So to solve 2×(5×q) + 16 = 37 we take the duplex of the 4 in (54q) from 37 and divide the result by 10 (twice the first digit again). This tells us that q must be 2 with 1 remaining:

$\sqrt{29_43_37_164} = 542$.

Now we see that the remaining two duplexes of 542 account for the remaining digits:
D(42) = 16, and D(2) = 4.

This explains the method given above and leads into the general square rooting method in which we use twice the first digit as a divisor repeatedly and reduce the dividend at each step by the duplex of all the answer figures after the first one.

14.3 GENERAL SQUARE ROOTS

Next we consider the general case where the square root does not terminate but has an infinite number of figures after the decimal point.
This is just an extension of the method above.

7 Find the first 5 figures and an approximate answer for the **square root of 38**.

There is clearly 1 figure before the point and it is a 6.
There is also a remainder of 2.
Then 20÷12=1 rem 8:

```
     3  8 . 0  0  0  0  0
12)        2  8
       6 . 1
```

From now on **we deduct the Duplex of all the figures** <u>**after the first**</u> (the 6) **from the diagonal figures** and then divide by 12.

So D(1)=1, 80–1=79, 79÷12=**6** rem **7**:

```
     3  8 . 0  0  0  0  0
12)        2  8  7
       6 . 1  6
```

Then D(16)=12, 70–12=58, 58÷12=4 rem 10:

```
     3  8 . 0  0  0  0  0
12)        2  8  7  10
       6 . 1  6  4
```

Then D(164)=44, 100–44=56, 56÷12=4 rem 8:

```
     3  8 . 0  0  0  0  0
12)        2  8  7  10  8
       6 . 1  6  4  4
```

So $\sqrt{38} \approx$ **6.1644**.

8 Find the first seven figures of $\sqrt{\mathbf{10330130}}$.

The procedure is just the same.
10'33'01'30 shows there are 4 figures before the point and the first one is 3:

```
     1  0  3  3  0  1  3  0 . 0
6)      1
        3              .
```

There is a remainder of 1 and the divisor is 6.

The full sum to seven figures is shown below:

```
      1  0  3  3  0  1  3  0.0
   6)      1  1  3  2  4  5  2
   ---------------------------
         3  2  1  4.0  5  2
```

The Duplexes to be found at each step are D(2), D(21), D(214), D(2140), D(21405).

Practice D Find to 5 significant figures the square root of:

a 27.2727 **b** 38.83 **c** 2929 **d** 11.23

e 3737 **f** 123356 **g** 707172 **h** 5000

i Find the first 9 figures of the square root of 26.123456789

j Find the first 8 figures of the square root of 17

a	**5.2223**	b	**6.2314**	c	**54.120**	d	**3.3511**
e	**61.131**	f	**351.22**	g	**840.94**	h	**70.711**
i	**5.11111111**	j	**4.1231056**				

As in the case of division sums it is sometimes necessary to alter an answer figure or to use bar numbers.

9 Find the **square root of 19.2**.

Initially we have:

```
    1  9.2  0  0
 8)       3
 ---------------
       4.
```

The next step is 32÷8=4 rem 0.
But this would mean subtracting 16 from 0 in the next step:

```
    1  9.2  0  0
 8)       3  0
 ---------------
       4.4
```

Method 1: Anticipating this happening we can avoid the negative numbers by saying 32÷8 = 3 rem 8 (see the second diagonal below) rather than 4 rem 0.

```
    1  9.2  0  0  0
 8)       3  8  7  14
 ---------------------
       4.3  8  1
```

Method 2: Alternatively we can accept the negative numbers:

$$\begin{array}{r|l} & \mathbf{1\ 9.2\ 0\ 0\ 0\ 0} \\ 8) & \quad 3\ \ 0\ \ 0\ \ 0\ \ 4 \\ \hline & 4.4\ \bar{2}\ \ 2\ \ \bar{3} \end{array} \qquad = \mathbf{4.3817}$$

After 32÷8 = 4 rem 0 in the second diagonal we have:

D(4) = 16, 0–16 = $\overline{16}$, $\overline{16}$÷8 = $\bar{2}$ rem 0.

Then D(4$\bar{2}$) = $\overline{16}$, 0–$\overline{16}$ = 16, 16÷8 = 2 rem 0.

And D(4$\bar{2}$2) = 20, 0–20 = $\overline{20}$, $\overline{20}$ ÷8 = $\bar{3}$ rem 4 (or $\bar{2}$ rem $\bar{4}$).

Sometimes it is convenient to introduce a bar number right at the beginning of a calculation, when the given number is just below a perfect square.

10 Find the first 5 figures of $\sqrt{34}$.

If we begin like this:

$$\begin{array}{r|l} & \mathbf{3\ 4.0\ 0\ 0} \\ 10) & \quad\ \ 9\ \ 10 \\ \hline & \ \ \mathbf{5.8} \end{array}$$

the large 8 here leads to large Duplex values and therefore frequent reduction of answer figures.

Alternatively, since 34 is close to 36, a perfect square, we could put 6 for the first figure and a remainder of $\bar{2}$:

$$\begin{array}{r|l} & \mathbf{3\ 4.0\ \ 0\ 0\ \ 0\ 0} \\ 12) & \quad \bar{2}\ \ 4\ \ 0\ \ 0\ \ \bar{5} \\ \hline & 6.\bar{2}\ \ 3\ \ 1\ \ 0 \end{array} \qquad = \mathbf{5.8310}$$

Practice E Find the first 4 figures in the square root of the following numbers (do at least the first 4 sums by both methods and remove any bar numbers from your answer).
For the last four use the method of Example 10 above.

a 27	**b** 39.6	**c** 1930	**d** 11.5
e 575	**f** 53	**g** 5	**h** 2
i 35	**j** 24	**k** 3	**l** 8.321

a 5.196	**b 6.292**	**c 43.93**	**d 3.391**
e 23.97	**f 7.280**	**g 2.236**	**h 1.414**
i 5.916	**j 4.899**	**k 1.732**	**l 2.884**

14.4 CHANGING THE DIVISOR

Because the first figure of a square root can take any whole value from 1 to 9, the divisor ranges from 2 to 18 inclusive.

Having 2 as a divisor can mean having to decide carefully about each answer digit so that the dividend at each step is kept manageable. Having 14, 16 or 18 as a divisor means mental division by these numbers.

11 Find $\sqrt{\mathbf{112.9}}$.

$$\begin{array}{r|llllll} & \mathbf{1} & \mathbf{1} & \mathbf{2} & .\mathbf{9} & \mathbf{0} & \mathbf{0} \\ 2) & {}_0 & {}_1 & & {}_0 & {}_3 & {}_4 \\ \hline & 1 & 0 & . & 6 & 3 & \bar{5} \end{array} = \mathbf{10.625}$$

$$\begin{array}{r|lllllll} & \mathbf{1} & \mathbf{1} & \mathbf{2} & .\mathbf{9} & \mathbf{0} & \mathbf{0} & \mathbf{0} \\ 20) & & & {}_{12} & {}_9 & {}_{14} & {}_{16} & \\ \hline & & 1 & 0 & .6 & 2 & 5 & \end{array}$$

In the first method above we divide by 2 and we need to think carefully about the appropriate answer figure and remainder at each step.

This can be alleviated however, as shown on the right, by looking at the first three figures in the number, 112, instead of only the first. That way we get 10 as the 'first figure', 20 as a divisor, and 12 as the remainder. This 10 is not involved in the duplex subtractions. Dividing by 20 is quite easy and there will be no reductions to consider.

12 Find $\sqrt{\mathbf{87.37}}$.

$$\begin{array}{r|lllll} & \mathbf{8} & \mathbf{7}.\mathbf{3} & \mathbf{7} & \mathbf{0} & \mathbf{0} \\ 18) & & {}_6 & {}_9 & {}_{16} & {}_{10} \\ \hline & & 9.3 & 4 & 7 & \end{array}$$

$$\begin{array}{r|lllll} & \mathbf{8} & \mathbf{7}.\mathbf{3} & \mathbf{7} & \mathbf{0} & \mathbf{0} \\ 20) & & {}_{\overline{13}} & {}_{13} & {}_8 & {}_{\bar{4}} \\ \hline & & 10.\bar{7} & 4 & 7 & \end{array}$$

The first method is straightforward except that we have to divide by 18 at each step. But if we take 10 as the first figure and a remainder of $\overline{13}$ we can arrange to have 20 as a divisor.

Another option is to divide the given number by 4, take the square root (which gives a divisor of 8 or 10) and multiply by 2 at the end.

It is worth experimenting to find the preferred method.

HEURISTIC PROOF

In finding a square root we are solving an equation of the form $x^2 = N$.
Suppose $x = a.bcde....$, that is, $x = a + 10^{-1}b + 10^{-2}c + 10^{-1}d + \ldots$ where a, b, c, d etc. take the integral values 0, 1, 2, 3, 4, 5, 6, 7, 8 or 9.

Then we have $(a.bcde....)^2 = N$.
Expanding the left-hand side in duplexes from left to right:

$$(a^2) + (2ab) + (2ac + b^2) + (2ad + 2bc) + (2ae + 2bd + c^2) + \ldots = N.$$

Here the brackets show successive duplexes and we assume each is a power of ten below the one before. N has to be exhausted by these bracketed quantities.

Notice that the first term in each bracket after the first is twice the first digit, a, multiplied by b, c, d, e etc., and that the other terms in those brackets are the successive duplexes of the answer digits ignoring the first answer digit, a.

This shows the method because we take the square of the first digit from N and the remainder (multiplied by 10) is then divided by 2a to give b.
Then from the remainder (multiplied by 10) we subtract b^2 and divide again by 2a to get c.
And so on.

Finding the square root of a number is equivalent to solving the equation $x^2 = c$. However this method can be extended to the solution of equations of the form $ax^2 + bx = c$ and to higher order polynomial equations: see Manual 3 and Reference 5.

LESSON 15
DIVISIBILITY

SUMMARY

15.1 Elementary Parts – small divisors are not dealt with in this book.
15.2 The Ekadhika – used for the following tests.
15.3 Osculation – the Vedic divisibility process.
15.4 Testing Longer Numbers – extension of osculation technique to numbers with many digits.
15.5 Other Divisors – testing for the factors of a potential divisor.
15.6 The Negative Osculator – a parallel osculator method involving divisors ending in 1.

15.1 ELEMENTARY PARTS

We can omit the earlier parts of this topic (as they are well-known) except to mention that *Only the Last Terms* is used to test for divisibility by 2, 4, 8, 5, 10. To test for divisibility by 3 and 9 we use *By Addition.*

So suppose we already know how to tell if a number is divisible by 2, 3, 4, 5, 6, 8, 9, 10 and numbers like 6 and 15 which can be expressed as a product of two or more numbers which are relatively prime.

Next we see how to test for divisibility by larger numbers, and especially prime numbers.

15.2 THE EKADHIKA

You will recall that the **Ekadhika** is the number "one more" than the one before. In this section the Ekadhika is the number *One More Than the One Before* when the number ends in a 9 or a series of 9's.

1 So for example for **19** the Ekadhika is **2** because the one before the 9 is 1 and one more than 1 is **2**.

For **69** the Ekadhika is **7**.

2 And for **13** the Ekadhika is **4** because **to obtain a 9 at the end we must multiply 13 by 3**, which gives 39, for which the Ekadhika is **4**.

Similarly, for **27**, to get a 9 at the end we multiply 27 by 7 which gives 189. So the Ekadhika for 27 is **19** (one more than 18 is 19).

Practice A Find the Ekadhika for each of the following numbers:

a 29	**b** 89	**c** 109	**d** 23	**e** 43	**f** 7	**g** 17	**h** 21

---

a 3 **b** 9 **c** 11 **d** 7 **e** 13 **f** 5 **g** 12 **h** 19

15.3 OSCULATION

There are two types of osculators: the positive osculator and the negative osculator.
The positive osculator is just the Ekadhika. We will deal with the negative osculator later.

A simple example will illustrate the osculation procedure, which follows the sub-Sutra *By Osculation.*.

3 Find out if **91** is divisible by **7**.

The Ekadhika for 7 is 5, **so we osculate the 91 with 5**.

> We osculate a number by multiplying its last figure by the osculator and adding the result to the previous figure.

This means we multiply the 1 in 91 by the osculator, 5, and add the result to the 9.

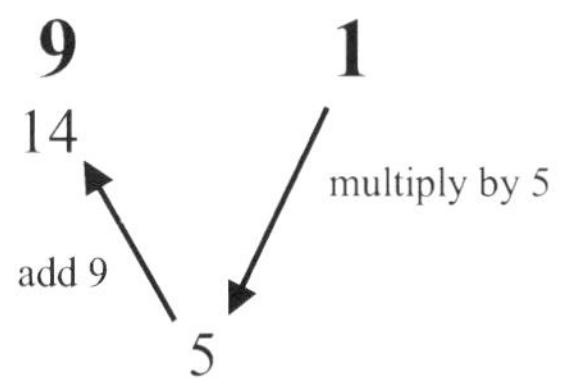

We get **14** as the result.
Since 14 is clearly divisible by 7 we can say that 91 is also divisible by 7.

The result we get from osculation (14 above) can also be osculated: in fact we can continue to osculate as many times as we like.

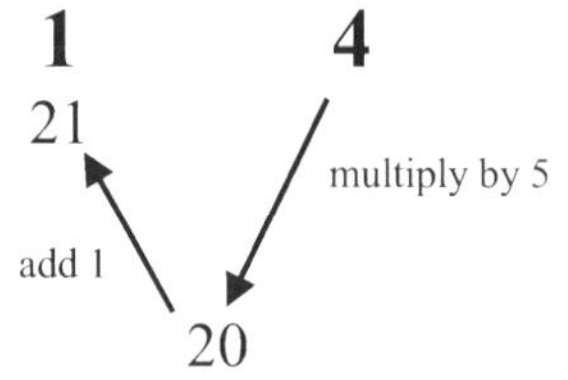

If we osculate **14** with 5 we get 4×5 + 1 = **21** (also clearly a multiple of 7).

If we osculate this result, **21**, we get 1×5 + 2 = **7**.

If we osculate **7** (think of **07**) we get 7×5 + 0 = **35**.

If you continue this osculation process you will see that only multiples of 7 are produced!

We find that osculating any multiple of 7 with the osculator, 5, always produces a multiple of 7. And osculating any number which is not a multiple of 7 will never produce a multiple of 7.

❋ Osculate 16 (which is not a multiple of 7) with the osculator, 5, and continue to osculate until you have produced at least 8 results. None of your results will be a multiple of 7.

Answers: 16, 31, 8, 40, 4, 20, 2, 10, 1, 5, 25, 27 . . .

The explanation for this is given at the end of this section.

4 Test **78** for divisibility by **13**.

First we find the Ekadhika for 13, which is 4 (E=4).

Then osculate 78 with this 4: 8×4 + 7 = **39**.
Since 39 is clearly a multiple of 13 we can say that 78 is divisible by 13.

If we did not recognise 39 as a multiple of 13 we could continue to osculate:
For 39: 9×4 + 3 = 39.
Here the 39 gets repeated and this also indicates divisibility.
Repetition indicates divisibility.

5 Test **86** for divisibility by **13**.

We osculate 86 with **4**: $6 \times 4 + 8 = 32$.
This is not a multiple of 13 so 86 is not divisible by 13.

But we can continue osculating if we do not see that 32 is not a multiple of 13.
Osculating the 32 we get $2 \times 4 + 3 = 11$.
Clearly 11 is not a multiple of 13, so 86 is not divisible by 13.

Practice B Test the following for divisibility by 7:

a 63 **b** 84 **c** 53 **d** 98

Test for divisibility by 13:

e 91 **f** 52 **g** 86 **h** 65

Test for divisibility by 19:

i 57 **j** 95 **k** 76 **l** 114 (multiply 4 by the osculator and add on 11)

Test for divisibility by 29:

m 87 **n** 116 **o** 57 **p** 58

Test for divisibility by 17:

q 85 **r** 56 **s** 51 **t** 102

a	yes	b	yes	c	no	d	yes
e	yes	f	yes	g	no	h	yes
i	yes	j	yes	k	yes	l	yes
m	yes	n	yes	o	no	p	yes
q	yes	r	no	s	yes	t	yes

EXPLANATION

First note the following two facts:
1. addition or subtraction of zeros to or from the end of a number do not affect the divisibility or non-divisibility of that number by the potential divisor;
2. addition or subtraction of the divisor or multiples of the divisor do not affect the divisibility or non-divisibility of a number.

So if we want to know if 247 is divisible by 19, which is the same as $2\bar{1}$, we add $2\bar{1}$ as many times as are needed to bring the 7 in 247 to zero, which means adding it 7 times.
This means we are adding $\bar{1}$ seven times to the 7 in 247 and adding 2 seven times to the 4 in 247.

This gives $2/_18/0$ and the zero can be dropped off to give $2/_18$.
We can then either repeat this process with the 18 (adding $2\bar{1}$ 18 times) or else we can carry the subscript 1 over to the left and remove the 8 by adding $2\bar{1}$ eight times.
In the latter case we add $16/\bar{8}$ to 3/8 to get 19/0 and dropping the zero leaves 19.

15.4 TESTING LONGER NUMBERS

If the number we are testing is a long number the osculation procedure is simply extended.

6 Is **247** divisible by **19**?

The Ekadhika is **2**.

We multiply the 7 by 2 and add on the next figure, the 4:

$$\begin{array}{ccc} \mathbf{2} & \mathbf{4} & \mathbf{7} \\ & \mathbf{18} & \end{array}$$

7×2 + 4 = 18. We put this under the 4 as shown.

We then multiply this 18 by 2 and add the 2 at the left of 247:

$$\begin{array}{ccc} \mathbf{2} & \mathbf{4} & \mathbf{7} \\ \mathbf{38} & \mathbf{18} & \end{array}$$

18×2 + 2 = 38 which is two 19's so **247 is divisible by 19**.

There is a useful short-cut however which prevents the numbers on the bottom line from getting too big.

Suppose the first stage is completed, so that we have:

$$\begin{array}{ccc} \mathbf{2} & \mathbf{4} & \mathbf{7} \\ & \mathbf{18} & \end{array}$$

Instead of multiplying 18 by 2 and adding the 2 on, we can suppose that the 1 in the 18 is carried over to join the 2. So we multiply only the 8 by 2 and add the 1 next to it *and* the 2.

So 8×2 + 1 + 2 = 19:

$$\begin{array}{ccc} \mathbf{2} & \mathbf{4} & \mathbf{7} \\ \mathbf{19} & \mathbf{18} & \end{array}$$

This is much easier and the 19 shows again that 247 is divisible by 19.

"These and many more interesting features there are in the Vedic decimal system, which can turn mathematics for the children from its present excruciatingly painful character to the exhilaratingly pleasant and even funny and delightful character it really bears ."
From "Vedic Mathematics", Page 239.

7 Is **4617** divisible by **19**?

The Ekadhika is still **2**.

We multiply the 7 by 2 and add on the 1:

4	**6**	**1**	**7**
		15	

Then multiply the 5 by 2 and add on the 1 and the 6:

4	**6**	**1**	**7**
	17	**15**	

Then multiply the 7 by 2 and add on the 1 and the 4:

4	**6**	**1**	**7**
19	**17**	**15**	

We end up with 19 and so **4617 is divisible by 19**.

Practice C Test the following numbers for divisibility by 19:

a 2774 **b** 589 **c** 323 **d** 4313 **e** 779 **f** 4503

g 14003 **h** 1995 **i** 10203 **j** 30201 **k** 234555

---

a yes b yes c yes d yes e yes f yes
g yes h yes i yes j no k yes

The Ekadhika is 2 for all these. Testing for divisibility for other numbers than 19 means using a different Ekadhika.

8 Is **13455** divisible by **23**?

The Ekadhika for 23 is **7**.

We simply osculate as before using this 7:

1	**3**	**4**	**5**	**5**
69	**59**	**8**	**40**	

We have 69 at the end which is clearly three 23's, so **yes: 13455 is divisible by 23**.

Practice D Test the following numbers for divisibility by the number shown:

a 41963 by 29 **b** 4802 by 49 **c** 4173 by 13 **d** 2254 by 23 **e** 10404 by 17

f 1003 by 59 **g** 4171 by 43 **h** 5432 by 7 **i** 4321 by 109

---

a yes b yes c yes d yes e yes
f yes g yes h yes i no

15.5 OTHER DIVISORS

9 Is **6308** divisible by **38**?

We see that 38 = 2×19 so we have to test for divisibility by 2 **and** 19.
The number is clearly divisible by 2 (because the last figure is even) so we only have to test for 19:

6	3	0	8
19	16	16	

6308 is also divisible by 19 so **it is divisible by 38**.

10 Is **5572** divisible by **21**?

Since 21 = 3×7 we must test for 3 and 7.
It is not divisible by 3 (digit sum is 1) so we need go no further:
5572 is not divisible by 21.

11 Is **1764** divisible by **28**?

We must test for 4 and 7 since 28 = 4×7 (note that 4 and 7 are relatively prime, we do not use 28 = 2×14 as 2 and 14 are not relatively prime).
The last two figures of 1764 indicate that it is divisible by 4 so we test for 7 next.

The osculator is 5:

1	7	6	4
49	39	26	

and we see that the test for 7 is passed.
So **1764 is divisible by 28**.

In testing for divisibility by a number we look at the factors of that number first and start with the easiest factors.

Practice E Test the following numbers for divisibility by the number shown:

a 3538 by 58 **b** 1254 by 38 **c** 21645 by 65

d 1771 by 46 **e** 767 by 95 **f** 5985 by 95

g 37932 by 58 **h** 334455 by 39 **i** 305448 by 52

---

a	yes	b	yes	c	yes
d	no	e	no	f	yes
g	yes	h	no	i	yes

If we want to test for divisibility by **31** we would have to multiply it by 9 to get a 9 in the last place. This would give 279 and an Ekadhika of **28**, which is too large to osculate with easily. So we need an alternative method here and this is where we use the negative osculator.

15.6 THE NEGATIVE OSCULATOR

12 The negative osculator for 31 is **3**, we just drop the 1.

The negative osculator for 41 is **4**. And so on.

To get the negative osculator for 17 we need to get a 1 at the end of the number and this can be done by multiplying 17 by 3. This gives 51 so the negative osculator for 17 is **5**.

Practice F Obtain the negative osculator for:

a 61	**b** 91	**c** 101	**d** 11	**e** 27	**f** 37
g 7	**h** 13	**i** 23	**j** 19		

---

a 6	**b** 9	**c** 10	**d** 1	**e** 8	**f** 11
g 2	**h** 9	**i** 16	**j** 17		

13 Is **3813** divisible by **31**?

We can use the negative osculator here, which is **3**.
The osculation process is slightly different for the negative osculator.
We begin by putting a bar over every other figure in 3813, starting with the second figure from the right:

$$\bar{3} \quad 8 \quad \bar{1} \quad 3$$

We then osculate as normal except that **any carry figure is counted as negative**.

$$\begin{array}{cccc} \bar{3} & 8 & \bar{1} & 3 \\ \mathbf{0} & \mathbf{32} & \mathbf{8} & \end{array}$$

$3 \times 3 + \bar{1} = \mathbf{8}$,
$8 \times 3 + 8 = \mathbf{32}$,
$2 \times 3 + \bar{3} + \bar{3} = \mathbf{0}$.

This zero indicates that **3813 is divisible by 31**.

14 Test **a 367164** and **b 6454** for divisibility by **7**?

Multiplying the 7 by 3 gives 21, so we can use a negative osculator of **2**.
a We put a bar on every other figure and osculate as above:

$$\begin{array}{cccccc} \bar{3} & 6 & \bar{7} & 1 & \bar{6} & 4 \\ \mathbf{0} & \mathbf{12} & \mathbf{3} & \mathbf{5} & \mathbf{2} & \end{array}$$

b Similarly:

$$\begin{array}{cccc} \bar{6} & 4 & \bar{5} & 4 \\ \bar{7} & \mathbf{10} & \mathbf{3} & \end{array}$$

Since $\bar{7}$ is a multiple of 7 we find that **both these numbers are divisible by 7**.

15 Is **11594** divisible by **62?**

Since 62 = 31×2 we need to test for 31 and 2.
The number is certainly divisible by 2 so we test for 31 by osculating with 3:

$$\begin{array}{ccccc} 1 & \bar{1} & 5 & \bar{9} & 4 \\ \mathbf{0} & \mathbf{10} & \mathbf{14} & \mathbf{3} & \end{array}$$

The number is divisible by 2 and 31 and is therefore **divisible by 62**.

Similarly, asked if the above number is divisible by 63, since 63 = 9×7 we would check for divisibility by 9 and then (if it passed this test) test for 7 by either using a positive osculator of 5 or a negative osculator of 2.

[We would not use 63 = 3×21 because we would be checking for divisibility by 3 twice and not testing for divisibility by 9. **The factors we split 63 into must be relatively prime**.]

Practice G Test the following for divisibility by the given number:

a 2914 by 31	**b** 9576 by 21	**c** 6039 by 61	**d** 20022 by 71
e 73472 by 41	**f** 63909 by 81	**g** 1728 by 91	**h** 7072 by 17
i 14715 by 27	**j** 7071 by 61	**k** 178467 by 31	**l** 45787 by 7
m 2394 by 42	**n** 4838 by 82	**o** 17949 by 93	**p** 9658 by 11

a **yes**	b **yes**	c **yes**	d **yes**
e **yes**	f **yes**	g **no**	h **yes**
i **yes**	j **no**	k **yes**	l **yes**
m **yes**	n **yes**	o **yes**	p **yes**

An interesting and useful result is that if you are testing for divisibility by a number, D, and its positive and negative osculators are P and Q respectively, then P + Q = D. So if we know P we can find Q and vice versa.

For further details on this topic see Manual 3 or References 1 and 3.

LESSON 16
COMBINED CALCULATIONS

SUMMARY

16.1 Algebraic – algebraic products in one line.
16.2 Arithmetic – finding sums of products etc.
16.3 Pythagoras Theorem – finding the square root of the sum of two squares in one line, digit by digit.

16.1 ALGEBRAIC

1 Expand and simplify: $\mathbf{(x + 3)(x + 4) + (x - 7)(x + 5)}$.

There is no need to do a lot of intermediate paper work when we can write the answer straight down.

There will clearly be two x^2 terms, one from each product, so we have $\mathbf{2x^2}$.
There will also be 7x from the first product and –2x from the second, giving **5x**.
And we also get +12 and –35, making **–23**.

So $(x + 3)(x + 4) + (x - 7)(x + 5) = \mathbf{2x^2 + 5x - 23}$.

2 Expand and simplify: $\mathbf{(2x - 3)^2 - (3x - 2)(x + 5)}$.

We have $4x^2 - 3x^2 = \mathbf{x^2}$.
Then $-12x - 13x = \mathbf{-25x}$.
And $9 - -10 = \mathbf{19}$.

So $(2x - 3)^2 - (3x - 2)(x + 5) = \mathbf{x^2 - 25x + 19}$.

With a little practice it becomes easy.

3 Expand: $\mathbf{(2x + 3y + 4)(x - 3y + 5)}$.

Here we can use the *Vertically and Crosswise* pattern from left to right to easily get the product term by term.

$$\begin{array}{ccccc} \mathbf{2x} & + & \mathbf{3y} & + & \mathbf{4} \\ \mathbf{x} & - & \mathbf{3y} & + & \mathbf{5} \\ \hline \multicolumn{5}{c}{\mathbf{2x^2 - 3xy - 9y^2 + 14x + 3y + 20}} \end{array}$$

Practice A Expand and simplify the following:

a $(x + 2)(x + 3) + (x + 1)(x + 5)$

b $(2x - 3)(x + 4) + (x - 1)(x + 3)$

c $(2x - 3)(x - 4) + (3x + 1)(x + 3)$

d $(2x + 3)(3x + 1) + (x + 3)(x + 4) + 5x + 6$

e $(x + 3)^2 + (x + 2)(x + 5)$

f $(x + 4)^2 + (x + 2)^2$

g $(5x + 2)^2 + (2x + 1)^2$

h $(5x + 2)^2 - (2x + 1)^2$

i $(x - 2)^2 + (2x - 1)^2$

j $(2x + 3)^2 + (x - 2)^2 + (x - 5)^2$

k $2x + (x + 1)^2 - 3$

l $(x + 2y + 3)(2x + y + 1)$

---

a	$2x^2 + 11x + 11$	b	$3x^2 + 7x - 15$
c	$5x^2 - x + 15$	d	$7x^2 + 23x + 21$
e	$2x^2 + 13x + 19$	f	$2x^2 + 12x + 20$
g	$29x^2 + 24x + 5$	h	$21x^2 + 16x + 3$
i	$5x^2 - 8x + 5$	j	$6x^2 - 2x + 38$
k	$x^2 + 4x - 2$	l	$2x^2 + 5xy + 2y^2 + 7x + 5y + 3$

4 Expand and simplify: $\mathbf{(x + 2)(x + 3)(x + 4)}$.

Here we expect a cubic (with an x^3-term, x^2-terms, x-terms and the final independent term).
Multiplying three brackets means **we multiply one term from each of the three brackets** in all possible ways.

Multiplying the first term in each bracket gives $x \times x \times x = \mathbf{x^3}$.
The x^2-term is formed by multiplying the x terms from two brackets and the independent term in the third. There are three ways of doing this, shown in bold below:

$(\mathbf{x} + 2)(\mathbf{x} + 3)(x + \mathbf{4})$ gives $\mathbf{x \times x \times 4 = 4x^2}$

$(x + \mathbf{2})(\mathbf{x} + 3)(\mathbf{x} + 4)$ gives $\mathbf{x \times x \times 2 = 2x^2}$

$(\mathbf{x} + 2)(x + \mathbf{3})(\mathbf{x} + 4)$ gives $\mathbf{x \times x \times 3 = 3x^2}$

So we find $\mathbf{9x^2}$ altogether.

For the x-term of the answer we multiply an x-term in one bracket by the independent terms in the other two brackets:

$(\mathbf{x} + 2)(x + \mathbf{3})(x + \mathbf{4})$ gives $\mathbf{3 \times 4 \times x = 12x}$

$(x + \mathbf{2})(\mathbf{x} + 3)(x + \mathbf{4})$ gives $\mathbf{2 \times 4 \times x = 8x}$

$(x + \mathbf{2})(x + \mathbf{3})(\mathbf{x} + 4)$ gives $\mathbf{2 \times 3 \times x = 6x}$

So we find $\mathbf{26x}$ altogether.

The independent term of the answer is just the product of the three independent terms:
2×3×4 = 24.

So our answer is $(x + 2)(x + 3)(x + 4) = \mathbf{x^3 + 9x^2 + 26x + 24}$.

5 Expand and simplify: $\mathbf{(2x + 3)^3}$.

We can apply the process shown above, by writing
$(2x + 3)^3$ as $(2x + 3)(2x + 3)(2x + 3)$ or we can use the expansion of $(a + b)^3$:

$$\mathbf{(a + b)^3 = a^3 + 3a^2b + 3ab^2 + b^3}$$

This means we cube the first term $(2x)^3$, which is $\mathbf{8x^3}$,
then find 3 times the first term squared times the second term: $3 \times (2x)^2 \times 3 = \mathbf{36x^2}$,
then 3 times the first term times the second term squared: $3 \times (2x) \times (3)^2 = \mathbf{54x}$,
and finally cube the second term: $(3)^3 = \mathbf{27}$.

So the answer is $(2x + 3)^3 = \mathbf{8x^3 + 36x^2 + 54x + 27}$.

✎ **Practice B** Use any one-line method to expand the following:

a $(x + 1)(x + 2)(x + 3)$ **b** $(x + 2)(x + 4)(x + 6)$

c $(2x + 1)(3x + 1)(x + 3)$ **d** $(x + 2)(x + 3)(x - 4)$

e $(x + 1)(3x - 1)(x - 3)$ **f** $(x + 1)(x + 3)(x + 5) - (x + 2)(x + 4)$

g $(x + 2)^3$ **h** $(2x + 1)^3$ **i** $(2x + 5)^3$

---

a $\mathbf{x^3 + 6x^2 + 11x + 6}$ b $\mathbf{x^3 + 12x^2 + 44x + 48}$
c $\mathbf{6x^3 + 23x^2 + 16x + 3}$ d $\mathbf{x^3 + x^2 - 14x - 24}$
e $\mathbf{3x^3 - 7x^2 - 7x + 3}$ f $\mathbf{x^3 + 8x^2 + 17x + 7}$
g $\mathbf{x^3 + 6x^2 + 12x + 8}$ h $\mathbf{8x^3 + 12x^2 + 6x + 1}$ i $\mathbf{8x^3 + 60x^2 + 150x + 125}$

16.2 ARITHMETIC

In the case of arithmetic calculations we have increased flexibility as we can choose when and how to introduce the vinculum, whether to work from left to right or right to left etc. The Sutra in use is *Vertically and Crosswise*.

The ability to work from left to right means that we can combine calculations and find sums of products, sums of squares etc. in one line. And if we want only the first 3 or 4 figures of an answer we save ourselves a lot of work by finding only those figures. In fact for some calculations there is no last figure to work from: for example there is no last figure for the square root of 2. We need to find these from left to right.

We deal here with sums of products and sums of squares. This can be developed further for evaluating trigonometric functions etc. (see Manual 3 or Reference 3).

Here we assume the left to right methods in Lessons 1, 6 and 7 have already been covered.

6 **$(53 \times 6) + (72 \times 4) = 6_{\bar{2}}\ 0\ 6$.**

To find the most significant figure (the left-most figure) we multiply the 5 by the 6 and the 7 by the 4 and add the two results: $30 + 28 = 58$, but the large digit (8) here suggests we put $6\bar{2}$, as shown.

We then multiply the 3 by 6 and the 2 by 4 to get 26.
To this we add the $\bar{2}$, as $\overline{20}$, to get 06, which we put down.

Had we put down 5_8 as the first result instead of $6_{\bar{2}}$ we would have had 26 again for the next step, and when these were combined we would get $26 + 80 = 106$, giving $5\ {}^{1}0\ 6 = 606$.

7 **$(293 \times 4) + (709 \times 6) = (3\bar{1}3 \times 4) + (71\bar{1} \times 6) = 5_4\ 4_2\ 2\ \ 6 = 5426$.**

We can use the vinculum here to change 293 to $3\bar{1}3$ and 709 to $71\bar{1}$.
It is probably better not to change the 7 into $1\bar{3}$ as it is easier to deal with two 3-figure numbers rather than one 3-figure and one 4-figure number.

Multiplying the left-most digits then gives $12+42 = 54$, which we put down.
The next pair of products gives $\bar{4}+6 = 2$ and combining this with the carried 4 (as 40) gives 42, which we put down.

The last pair of products gives $12+\bar{6} = 6$ and adding the carry we get 26 which we put down.

It is worth noting that it is often best not to introduce too many vinculums. We want the positive and negative parts to cancel in the calculation, as far as possible, and so we need to leave some positive digits to help with this.

8 **$(6345 \times 4) + (257 \times 3) + (67 \times 8) = 2_4\, 6_{\bar{2}}\, 6_{\bar{1}}\, 8\; 7 = 26687$.**

Here we have three products.
We begin by finding the number of thousands and for this we only need to look at the first product: $6 \times 4 = 24$, we put down 2_4.
Next, for hundreds we have 3×4 in the first product and also 2×3 in the second. This gives 18 altogether and with the carry 4 we get 58 which we write as $6_{\bar{2}}$.

Now for the tens we bring in all three products: $4 \times 4 + 5 \times 3 + 6 \times 8 = 79$.
To this we add the carried $\overline{20}$ to get 59, which we write as $6_{\bar{1}}$.

Lastly for the units we have 20+21+56 = 97. The carry makes this 87 which we put down.

9 **$(62 \times 6) - (38 \times 4) = 2_4\, 2\;\; 0 = 220$.**

Here we have to subtract the products.
The two products on the left are 36, 12 and subtracting gives 24 which we put down.

The other products give 12 – 32 = –20, and to this we add the carry: –20 + 40 = 20, which we put down.

✐ Practice C

a $(62 \times 6) + (38 \times 4)$

b $(43 \times 5) + (36 \times 3)$

c $(33 \times 4) + (35 \times 3) + (42 \times 6)$

d $(72 \times 4) - (34 \times 6)$

e $(83 \times 4) - (44 \times 6)$

f $(443 \times 4) + (345 \times 6)$

g $(513 \times 5) + (215 \times 4)$

h $(536 \times 6) + (405 \times 4) + (721 \times 3)$

i $(333 \times 7) + (34 \times 6)$

j $(4532 \times 4) + (34 \times 6)$

k $(7124 \times 6) + (192 \times 7)$

---

a	524	b	323
c	489	d	84
e	68	f	3842
g	3425	h	6999
i	2535	j	18332
k	44088		

"whatever is consistent with right reasoning should be accepted, even though it comes from a boy or even from a parrot; and whatever is inconsistent therewith ought to be rejected, although emanating from an old man or even from the great sage Shree Shuka himself".
quoted in "Vedic Mathematics", Page 1d.

10 **(36 × 49) + (62 × 42) = (36 × 5$\bar{1}$) + (62 × 42) = $4_{\bar{1}}3_7 6\ 8$ = 4368.**

First remove the 9 in 49.
Then the first product in each part is 3 × 5 and 6 × 4.
This gives 39 for the hundreds which we put down as $4_{\bar{1}}$.

Next we take the crosswise step in each part and add them: (30+$\bar{3}$) + (8+12) = 47.
With the carry this becomes 37 which we put down.

Finally for the units we get $\bar{6} + 4 = \bar{2}$.
Adding 70 to this gives 68 which we put down.

11 **$57^2 + 73^2 = 7_4\ {}^{1}5_2\ 7\ \ 8$ = 8578.**

Taking the duplex of the left-most digit in each number first we get:
D(5) + D(7) = 25+49 = 74, which we put down.

Then the next duplex: D(57) + D(73) = 70+42 = 112.
To this we add the carried 4 to get 152, as shown.
The 2-figure number we put down (15) means that the 1 will have to be carried leftwards (to the 7).

Next D(7) + D(3) = 58. And 58 + 20 = 78 which we put down.

Finally take the 1 in 15 over to the 7 to get 8578.

Other ways of doing this calculation are:

a) to put $8_{\bar{6}}$ down at the outset instead of 7_4. This gives $8_{\bar{6}}5_2 7\ 8$.

b) to change 57^2 to $6\bar{3}^2$. This gives $8_5\ 5_6\ 7\ 8$.

12 **(34321 × 62) + 413^2 = $2_{\bar{2}}2_6 9_4 8_3 4_6 7\ \ 1$ = 2298471.**

The first result in the bracket is 3 × 6 = 18, and this is actually followed by five zeros. The first result from 413^2 will be 16 followed by four zeros.

This means that the square does not come in until the second step.

So first we get 18, written as $2_{\bar{2}}$.
Next imagine 62 under 34:

```
3 4 3 2 1
6 2
```

Then crosswise gives 3 ×2 + 4 × 6 = 30. And to this we add the first duplex from 413^2, which we saw above was 16.

This gives 46 and we have a carry which makes this 26 which we put down.

The next crosswise step

```
3 4 3 2 1
  6 2
```

gives 26 and the next duplex is D(41) = 8. Adding these gives 34.
With the carry this now gives 94, put down 9_4.

Next we get (3 × 2) + (2 × 6) + D(413) = 18 + 25 = 43.
Adding the carry we get 8_3.

Then 10 + D(13) = 16. And 16 + carry = 46.
Finally 2 + D(3) = 11. And 11 + 60 = 71.

So we see that the method is quite straightforward: we combine all products of the highest order of magnitude first, then all those of the next order, and so on. We make full use of the vinculum, where appropriate, either changing the given numbers into vinculum form or changing results obtained during the calculation.

It is also wise to have an eye on the next stage of the calculation when deciding how to express the result of a particular stage.

✐ Practice D

a (34 × 44) + (53 × 63) **b** (48 × 42) + (72 × 82) **c** (81 × 26) + (54 × 48)

d $47^2 + 63^2$ **e** $53^2 + 72^2$ **f** $76^2 + 44^2$

g $42^2 + 83^2 + 56^2$ **h** $29^2 + 37^2 + 68^2$ **i** $52^2 + 63^2 + 28^2 + 34^2$

j $84^2 - 56^2$ **k** $49^2 + 53^2 - 66^2$ **l** (76 × 54) – (32 × 24)

m (61 × 53) – (54 × 43) **n** (66 × 32) + (57 × 75) **o** (444 × 32) + (617 × 8)

p $(64 \times 34) + 48^2$ **q** (4343 × 32) + (56 × 44) **r** $(479 \times 32) + 732^2$

Find to 3 S.F.

s (3749 × 27) + (187 × 75) **t** $7359^2 + 427^2$ **u** $(888.34 \times 73) - 187^2$

---

a	4835	b	7920	c	4698
d	6178	e	7993	f	7712
g	11789	h	6834	i	8613
j	3920	k	854	l	3336
m	911	n	6387	o	19144
p	4480	q	141440	r	551152
s	115000	t	54300000	u	29900

As an illustration of an application of combining operations we show finally how to evaluate $\sqrt{a^2+b^2}$, given a and b.

16.3 PYTHAGORAS THEOREM

In finding the hypotenuse of a right-angled triangle whose other sides are, say, 4 and 5, we need to find $\sqrt{4^2+5^2}=\sqrt{41}$. This just involves finding a square root, though, which has been covered in Lesson 14.

In finding $\sqrt{317^2+421^2}$ however we can proceed digit by digit by first finding the total of the duplexes of 3 and 4 and the first digit of the square root of the result, and then proceed to the next order down and so on.

13 Find $\sqrt{\mathbf{317^2+421^2}}$ to 4 significant figures.

We will take the duplexes of the two numbers from left to right, add them up, and then take the square root:

Writing the duplexes out we have: 9/6/43/14/49 + 16/16/12/4/1*
= 25/22/55/18/50.

Then for the square root we have:

$$\begin{array}{rl} & 25/22/55/18/50 \\ 10) & {}_{0}\quad {}_{2}\quad {}_{1}\quad {}_{0}\quad {}_{1} \\ \hline & \underline{5\ \ 2\ \ 7\ .\ 0\ \ 0}\ ... \end{array}$$

The answer is 527.0.

The first figure is clearly 5, rem 0; then 10 into 22 is 2 rem 2.
Then $75-2^2=71$, 10 into 71 is 7 rem 1.
Then $28-28=0$, 10 into 0 is 0 rem 0.
Then $50-49=1$, 10 into 1 is 0 rem 1.
Etc.

*Note that although we have calculated all the duplexes here, before finding the square root, in practice we need only find each duplex as we need it, add it to the other duplex, and obtain the next figure of the square root. See the next example.

14 Find $\sqrt{\mathbf{53.6^2 + 732^2}}$ to 4 significant figures

$$\begin{array}{rl} & 49/42/62/42/73/36 \\ 14) & \quad {}_{0\ \ 0\ \ \bar{3}\ \ 2\ \ \bar{1}} \\ \hline & 7\ \ 3\ \ 4\ .\ \bar{1}\ \ 6\ldots \\ \hline \end{array}$$

53.6 will not contribute to either of the first two duplexes from 732, so we begin with D(7) = 49, the square root is 7, there is no remainder, and the divisor is 14.

The next duplex is D(73) = 42, 14 into 42 is 3 rem 0.
Then D(5) + D(732) = 62, 62 – D(3) = 53, 14 into 53 is 4 rem $\bar{3}$.
[the D(3) here comes from the second answer digit.]
Then D(53) + D(32) = 42, 42 + $\overline{30}$ = 12, 12 – D(34) = $\overline{12}$, 14 into $\overline{12}$ is $\bar{1}$ rem 2.
Then D(536) + D(2) = 73, 73 + 20 = 93, 93 – D(34$\bar{1}$) = 83, 14 into 83 is 6 rem $\bar{1}$.
So the answer is 734.0.

There is no question of attempting this with conventional maths without a calculator or tables. Two long multiplications would be required, then an addition and finally a square root.

15 Find $\sqrt{\mathbf{69^2 - 43^2}}$ to 4 significant figures.

Here we will write 69 as $7\bar{1}$ and subtract the duplexes.
We will also abbreviate the working further.

$$\begin{array}{l} \qquad\qquad\qquad 10) \\ \sqrt{\mathbf{7\bar{1}^2 - 43^2}} = 5_8 4_2 . 0_{\bar{4}} \bar{4}_0 3_2 \ldots \end{array}$$

First $7^2 – 4^2 = 33$, so the first figure is 5, the first remainder is 8 and the divisor will be 10:

$$\begin{array}{r|l} 10) & \\ \hline & 5_8 \\ \hline \end{array}$$

Then the duplexes are: $2(\bar{7} – 12) = –38$, and 80 – 38 = 42 and 42 ÷ 10 = 4_2:

$$\begin{array}{r|l} 10) & \\ \hline & 5_8 4_2 . \\ \hline \end{array}$$

Next: duplexes, $(\bar{1})^2 - 3^2 = –8$; 20 – 8 = 12, $12 – 4^2 = –4$, –4 ÷ 10 = $0_{\bar{4}}$:

$$\begin{array}{r|l} 10) & \\ \hline & 5_8 4_2 . 0_{\bar{4}} \\ \hline \end{array}$$

Now we continue just like ordinary square roots a s there are no more duplexes to bring in: D(40) = 0, $\overline{40} - 0 = \overline{40}$, $\overline{40} \div 10 = \bar{4}_0$ and so on.

$$\begin{array}{r|l} 10) & \\ \hline & 5_8 4_2 . 0_{\bar{4}} \bar{4}_0 3_2 \ldots \\ \hline \end{array}$$

Practice E Evaluate the following, to 4 S.F.

a $\sqrt{33^2+41^2}$ **b** $\sqrt{23^2+34^2}$ **c** $\sqrt{81^2+21^2}$

d $\sqrt{81^2-21^2}$ **e** $\sqrt{302^2+411^2}$ **f** $\sqrt{512^2+33^2}$

g $\sqrt{613^2-321^2}$

---

a	**$5_0 2_6 . 6_6 3_6$**	b	**41.05**	c	**83.68**
d	**78.23**	e	**510.0**	f	**513.1**
g	**522.2**				

Calculating from left to right and combining calculations like this can be taken much further, as shown in Manual 3.

There are many other applications of this important *Vertically and Crosswise* sutra including calculation of trigonometric functions and their inverses, solution of polynomial and transcendental equations, matrices, triple arithmetic, solution of Diophantine equations and many applications in calculus (see Reference 5 and Reference 4).

"And as regards the time required by the students for mastering the whole course of Vedic Mathematics as applied to all its branches, we need merely state from our actual experience that 8 months (or 12 months) at an average rate of 2 or 3 hours per day should suffice for completing the whole course of mathematical studies on these Vedic lines instead of 15 or 20 years required according to the existing systems of Indian and also of foreign universities."
From "Vedic Mathematics", Page xvii.

VEDIC MATHEMATICS SUTRAS

1	एकाधिकेन पूर्वेन Ekādhikena Pūrveṇa	*By One More than the One Before*
2	निखिलं नवतश्चरमं दशतः Nikhilaṃ Navatascaramaṃ Dasataḥ	*All from 9 and the Last from 10*
3	ऊर्ध्वतिर्यग्भ्यामं Ūrdhva Tiryagbhyāṃ	*Vertically and Crosswise*
4	परावर्त्य योजयेत् Parāvartya Yojayet	*Transpose and Apply*
5	शून्यं साम्यसमुच्चये Sūnyaṃ Sāmyasamuccaya	*If the Samuccaya is the Same it is Zero*
6	आनुरूप्ये शून्यं अन्यत् (Ānurūpya) Sūnyamanyat	*If One is in Ratio the Other is Zero*
7	संकलन व्यवकलनाभ्यां Saṅkalana Vyavakalanābhyām	*By Addition and by Subtraction*
8	पूरणापूरणाभ्यां Pūraṇāpūraṇābhyām	*By the Completion or Non-Completion*
9	चलनकलनाभ्याम् Calana Kalanābhyām	*Differential Calculus*
10	यावदूनं Yāvadūnam	*By the Deficiency*
11	व्यष्टिसमष्टिः Vyaṣṭisamaṣṭiḥ	*Specific and General*
12	शेषाण्यङ्केन चरमेण Seṣaṇyaṅkena Carameṇa	*The Remainders by the Last Digit*
13	सोपान्त्यद्वयमन्त्यं Sopāntyadvayamantyam	*The Ultimate and Twice the Penultimate*
14	एकन्यूनेन पूर्वेन Ekanūnena Pūrveṇa	*By One Less than the One Before*
15	गुणितसमुच्चयः Guṇitsamuccayaḥ	*The Product of the Sum*
16	गुणकसमुच्चयः Guṇakasamuccayaḥ	*All the Multipliers*

SUB-SUTRAS

1	आनुरूप्येण Ānurūpyeṇa	*Proportionately*
2	शिष्यते शेषसंज्ञः Sisyate Seṣaṃjñaḥ	*The Remainder Remains Constant*
3	आधमाधेनान्त्यमन्त्येन Ādyamādyenāntyamantyena	*The First by the First and the Last by the Last*
4	केवलैः सप्तकं गुणयात् Kevalaiḥ Saptakaṃ Guṇyāt	*For 7 the Multiplicand is 143*
5	वेष्टनम् Veṣṭanaṃ	*By Osculation*
6	यावदूनं तावदुनं Yāvadūnaṃ Tāvadūnaṃ	*Lessen by the Deficiency*
7	यावदूनं तावदूनीकत्य वर्गं च योजयेत् Yāvadūnaṃ Tāvadūn ikṛtya Vargañca Yojayet	*Whatever the Deficiency lessen by that amount and set up the Square of the Deficiency*
8	अन्त्ययोर्दशकेऽपि Antyayordasake'pi	*Last Totalling 10*
9	अन्त्ययोरेव Antyayoreva	*Only the Last Terms*
10	समुच्चयगुणितः Samuccayagunitaḥ	*The Sum of the Products*
11	लोपनस्थापनाभ्यां Lopanasthāpanābhyām	*By Alternate Elimination and Retention*
12	विलोकनं Vilokanam	*By Mere Observation*
13	गुणितसमच्चुयः समुच्चयगुणितः Guṇitsamuccayaḥ Samuccayaguṇitaḥ	*The Product of the Sum is the Sum of the Products*
14	ध्वजाङ्क Dhvajāṅka	*On the Flag*

9-POINT CIRCLES

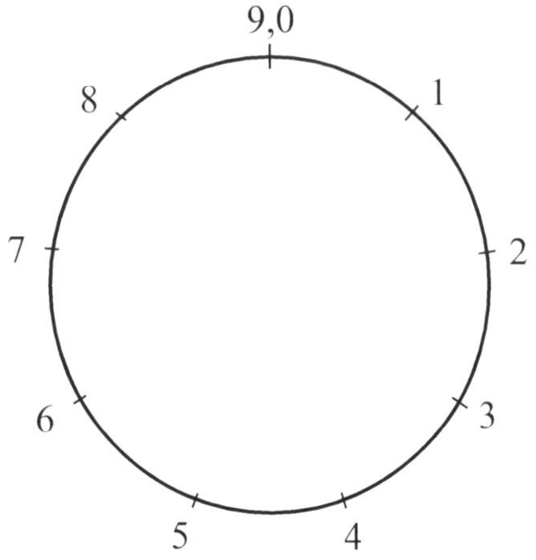

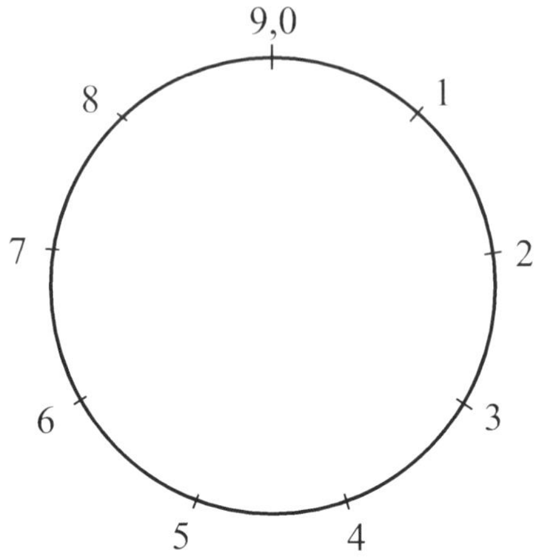

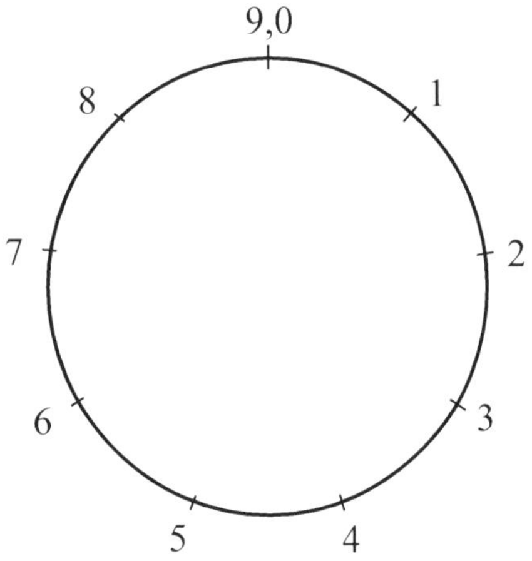

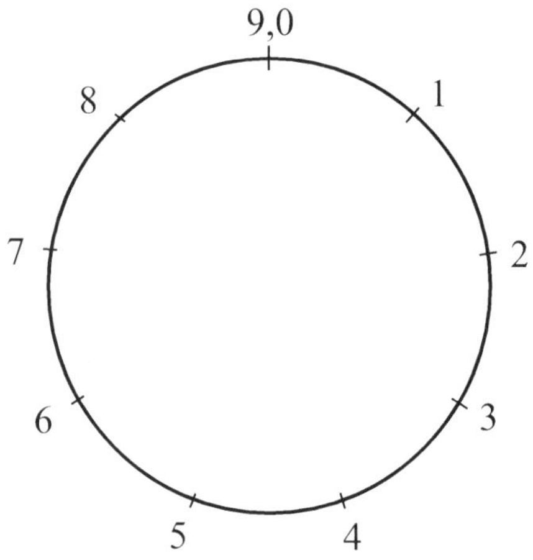

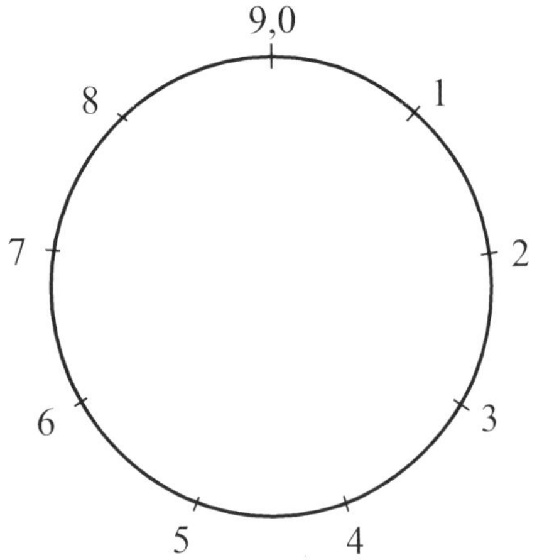

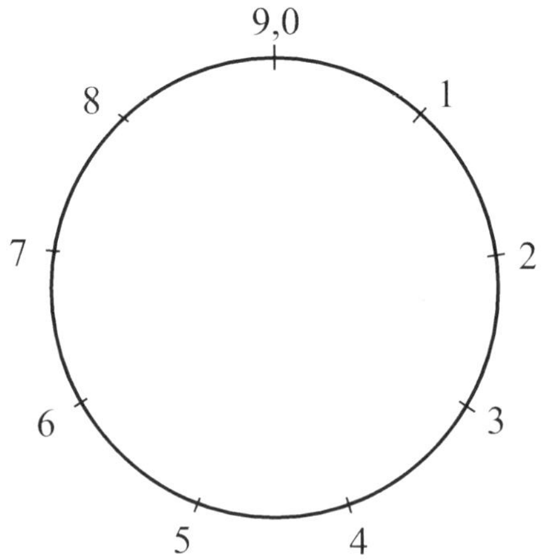

REFERENCES

1. Sri Bharati Krsna Tirthaji, "Vedic Mathematics", published by Motilal Banarsidass, 1965. ISBN 81-208-0163-6.

2. "Celebrating Perfection in Education", Maharishi University Press 1997. ISBN 81-7523-013-4.

3. Williams K. R. *Discover Vedic Mathematics*. Vedic Mathematics Research Group, 1984. ISBN 1-869932-01-3.

4. Williams K.R. *Triples*. Inspiration Books, 1999. ISBN 1-902517-00-8.

5. Nicholas A. P., K. Williams, J. Pickles. *Vertically and Crosswise*. Inspiration Books, 1984. ISBN 1-902517-03-2.

6. Sri Bharati Krsna Tirthaji, "Vedic Metaphysics", published by Motilal Banarsidass, 1978. ISBN 0-89581-472-2.

INDEX OF THE VEDIC FORMULAE

SUTRAS

INDEX OF THE VEDIC FORMULAE

SUB-SUTRAS

INDEX

INDEX

OTHER VEDIC MATHEMATICS BOOKS

Teacher's Manual – Elementary
This is similar to the intermediate book but covers a lesser range (aimed at teachers of grades 3 to 7). It contains many topics that are not in the other Manuals that are suitable for this age range and many topics that are also in Manual 2 are covered in greater detail here. ISBN 978-1-902517-16-2

Teacher's Manual – Advanced
This includes: calculus, series, logs and exponentials, trigonometry (including solving trig equations, proving identities), solution of equations (special types, quadratics, cubics, transcendental), complex numbers, coordinate geometry, transformation geometry, Simple Harmonic Motion, projectile motion, forces, work moments, etc. ISBN 978-1-902517-18-6

The Cosmic Computer - Abridged **Edition**
This is a shortened version of the full course described next. It contains some of the most striking Vedic methods. It is beautifully illustrated with full colour cover and is perfect bound (paperback). A good introduction to Vedic Mathematics. Authors: Kenneth Williams and Mark Gaskell. 216 pages. Almost A4 in size. 1997. ISBN 978-0-9531782-0-9

The Cosmic Computer Course
This covers Key Stage 3 (age 11-14 years) of the National Curriculum for England and Wales. It consists of three textbooks, a Teacher's Guide and an Answer Book. Much of the material in Book 1 is suitable for children as young as eight and this is developed from here to topics such as Pythagoras' Theorem and Quadratic Equations in Book 3. The Teacher's Guide contains a Summary of each Book, a Unified Field Chart (showing the whole subject of mathematics and how each of the parts are related), hundreds of Mental Tests (these revise previous work, introduce new ideas and are carefully correlated with the rest of the course), Extension Sheets (about 16 per book) for fast pupils or for extra classwork, Revision Tests, Games, Worksheets etc.

Discover Vedic Mathematics
This has sixteen chapters each of which focuses on one of the Vedic Sutras or sub-Sutras and shows many applications of each. It goes into some detail, showing variations and giving explanations and proofs.. Also contains solutions to GCSE and 'A' level examination questions. Author: K. Williams, 216 pages. 1984.

Triples
This book shows applications of Pythagorean Triples (like 3,4,5). A simple, elegant system for combining these triples gives unexpected and powerful general methods for solving a wide range of mathematical problems, with far less effort than conventional methods use. The easy text fully explains this method which has applications in trigonometry (you do not need any of those complicated formulae), coordinate geometry (2 and 3 dimensions) transformations (2 and 3 dimensions), simple harmonic motion, astronomy etc. etc. Author: K. Williams, 176 pages, paperback.

Vertically and Crosswise
This is an advanced book of sixteen chapters on one Sutra ranging from elementary multiplication etc. to the solution of non-linear partial differential equations. It deals with (i) calculation of common functions and their series expansions, and (ii) the solution of equations, starting with simultaneous equations and moving on to algebraic, transcendental and differential equations. Authors: A. P. Nicholas, K. Williams, J. Pickles, 200 pages, paperback, 1999.

The Natural Calculator
This book focuses on mental mathematics and has a detailed introduction outlining its merits. There are nine chapters, mainly on multiplication but including addition, subtraction and division. Author: K. Williams, 100 pages, paperback. ISBN 978-1-902517-15-5

For further details and other books please visit **http://www.vedicmaths.org**

Made in the USA
Monee, IL
04 November 2026

67375811R00122